Seed Quality

Dr. Gurnam Singh, a Ph.D. from Narendra Deva University of Agriculture and Technology, Faizabad, has worked with Seed Certification wing of UP State Agriculture Department in various capacities. He is a fellow of the two premier science academics of India and member of several professional societies. He has published 17 research and review papers in internationally reputed journals, besides five books on various aspects of seed science. He visited many countries, attended seminars, workshops and corporate conferences of National and International status.

This book has been designed as a referral for agriculturists and farmers to guide them on choosing the best available quality seeds. It focuses on instructing readers on how to choose superior quality certified seedsprogeny of seeds that have been approved by a certifying agency for their genetic identity and purity. A lot of emphasis has been placed upon various factors like the presence of weeds, dormancy, germination, winter and spring varieties, moisture content, etc. Also, tests which can be conducted to ascertain quality, such as viability and vigour tests, have also been explained.

All in all, this is a comprehensive guide exploring the different facets of seed quality. Equipping the readers which highly insightful and key information, it is hoped that this book proves to be highly resourceful.

Seed Quality

Editor

Gurnam Singh

2018

Gene-Tech Books

New Delhi - 110 002

ISBN10 81-89729-91-8
ISBN13 978-81-89729-91-2

Published by : **GENE-TECH BOOKS**
4762-63/23, Ansari Road, Darya Ganj,
NEW DELHI - 110 002
Phone: 41562849
e-mail: genetechbooks@yahoo.co.in

Printed at : **Tarun Offset Printers**
Delhi

PRINTED IN INDIA

Preface

The agricultural world today is a complex, technology - driven arena where each contributing factor is a vital input to the final picture. With seed science and technology being just such an input growing by leaps and bounds each passing year, it comes as no surprise that seed quality is of crucial importance in agriculture today.

This book has been designed as a referral for agriculturists and farmers to guide them on choosing the best available quality seeds. It focuses on instructing readers on how to choose superior quality certified seeds - progeny of seeds that have been approved by a certifying agency for their genetic identity and purity. A lot of emphasis has been placed upon various factors like the presence of weeds, dormancy, germination, winter and spring varieties, moisture content, etc. Also, tests which can be conducted to ascertain quality, such as viability and vigour tests, have also been explained.

All in all, this is a comprehensive guide exploring the different facets of seed quality. Equipping the readers which highly insightful and key information, it is hoped that this book proves to be highly resourceful.

Gurnam Singh

Contents

1

Seed Selection and Quality

Good quality seeds are essential to grow a strong and healthy crop. Healthy seeds can be bought from trusted sources or farmers can produce their own seeds. In that case, seed selection can be used to improve the quality of seeds.

There are several diseases that are transmitted via the seeds. If seeds from an infested field are used to grow the next crop, these seed-born diseases will immediately cause serious problems. Seed selection should thus start by obtaining seeds from healthy plants.

Small, shriveled and broken seeds contain less nutrition for the developing seedling. By removing these inferior seeds, the farmer is able to grow stronger and healthier seedlings.

Even if seeds are selected carefully, they still may carry diseases. In these cases seed treatment (e.g. hot water treatment or chemical seed treatment) could be used to further improve the quality of the seeds.

While seed selection is mainly aimed at obtaining healthier seeds, it can be used also to maintain and improve the quality of the crop variety. In a crop field, there are always differences between plants. Some plants may have characteristics that are more suitable than those of other plants. During the growing season, the farmer can try to observe these differences and mark preferred plants with a ribbon or with a stick. During the harvest, the seeds of these plants can be reserved for growing the next crop. In this way, the farmer can slowly improve the quality of his variety.

The selection of these plants may be based on characteristics such as the size of the plant, color or size of fruits, number of grains per ear, etc. But selection can also be done to keep seeds of plants that suffered less attacks by insects or diseases.

In this way, the farmer will select at harvest time the best seeds to be kept apart for the next season. Still, just before starting the next season, it is recommended to select once more to remove seeds that are too small, spotted, deformed,. discolored, etc.

For sowing only the very best seeds should be used.

Forage Seed Quality

Forage seed quality depends on three key components:

— germination,

— genetic purity, and

— crop purity.

Germination

Germination is the measure of how well a seed produces a normal, healthy seedling when planted under ideal conditions. The standard germination test, evaluated under the temperature considered ideal for each species, provides an estimation of seedling emergence if soil conditions at planting and during early seedling growth are near optimum. The results of the standard germination test are printed on the seed tag.

Each species has a specific range of temperatures at which germination occurs. If temperature at planting falls outside this range, germination may be slowed or prevented, and during extreme conditions the seedlings may die. Excessively high or low temperatures also can induce secondary seed dormancy, causing a delay in germination and establishment. It is therefore important for growers to be aware of the germination requirements of the species that is planted and to sow fields when conditions are favourable.

Alfalfa

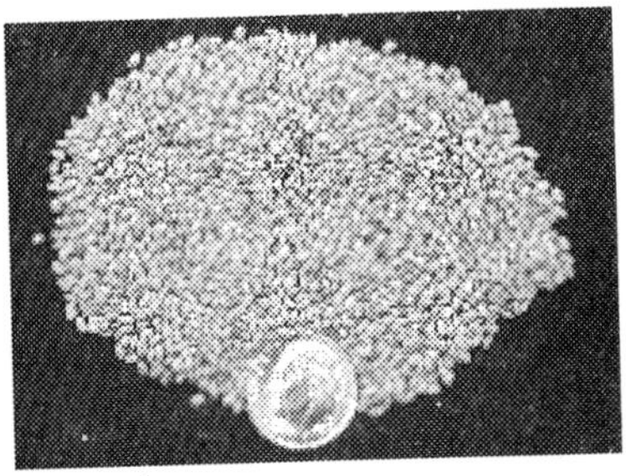

Clover

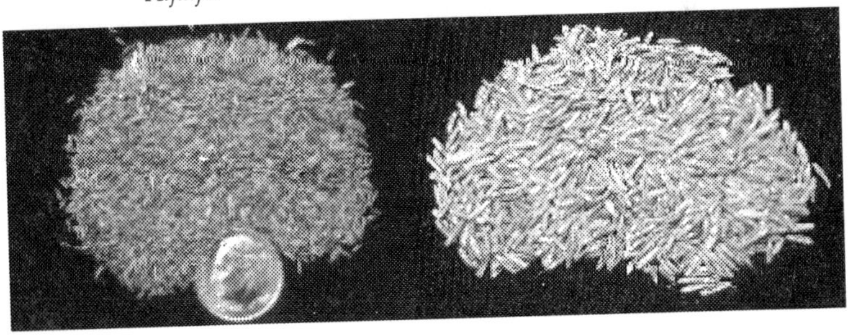

Fescue *Ryegrass*

It is essential that forage seed have adequate moisture immediately following planting to promote rapid germination and for a period of about three weeks after emergence to promote seedling growth.

Genetic Purity

Varieties are adapted for specific uses and production areas. The genetic constitution of a variety will influence disease and insect resistance, plant response to climatic extremes, forage quality, and many agronomic characteristics. Growers should purchase seed of a known variety that has been tested for their area.

In North Carolina, forage seed may be sold variety not stated. However, growers should realise that purchasing seed of an unknown variety often leads to low yields, reduced stand life, and increased cost.

Genetic purity, or trueness to variety, is established and maintained by special purification and seed increase programs, by field and seed inspections, and by pedigree records. The best assurance of obtaining genetically pure seed is to buy certified seed, clearly identified by the "blue tag". The blue certified seed tag is issued only to seed lots that meet specific genetic, physical, and physiological standards.

Crop Purity

Seed lots with even small amounts of weed seed or other crop seed can cause serious economic losses and diminish forage quality. Weeds and other crop plants compete with the desired species for nutrients, space, and soil moisture and are often difficult or expensive to control. Some weeds, such as serrated tussock, Canada thistle, or horsenettle, can harm animals or threaten pasture value. Other weeds that are frequently found in forage seed include dock, sedge, wild garlic, wild onion, mustards, panicums, and other forage seeds such as ryegrass.

Seed Analysis Tag

The North Carolina Seed Law requires containers of seed

offered for sale to be labeled for content. The seed analysis tag or label contains information based on germination and purity analysis of a representative sample taken from the seed lot.

The N.C. Seed Law prohibits the sale of forage seed that:

— Fall below 70 percent germination.

— Contain more than 1 percent weed seed.

— Contain excessive amounts of restricted noxious weed seed.

— Contain any prohibited noxious weed seed.

Here is a description of the information offered on the seed tag:

Kind and variety

The "kind" of crop refers to the species. North Carolina does not require the forage variety name to be listed on the tag. However, if the variety is not given, the seed tag should bear the statement, variety not stated. When two or more varieties or kinds are named, the word "mixture" or "mixed" must appear on the label. Bermudagrass, often called common bermuda, is no exception to these rules. The sale of "common" bermudagrass requires either the variety name or the notation variety not stated.

Germination

This notation tells the percentage of pure seed that germinates. The percentage is based on the number of seeds that produced normal seedlings following a specified test period (a normal seedling being one that has the essential structures necessary for plant survival). If more than one kind and variety are named, the germination percentage must be shown for each.

Some forage species are dormant when harvested, and the dormancy period may last for a few weeks to several months. When seeds are dormant, they are still viable but will not germinate when placed in the proper germination environment. Special pre-conditioning treatments must be used to break or overcome this dormancy. Switchgrass is a good example. Its seeds must have a pre-chill treatment for 2 weeks at 40°F to condition the seed for germination. Also, the seeds must be fully imbibed during the pre-chill treatment. When switchgrass is labeled 90 percent germination on the seed tag, it takes into account two weeks' pre-chill treatment during the test to break dormancy, followed by two weeks in alternating 60/85°F to measure germination potential.

It is possible to buy seeds that have high germination potential, but if dormancy has not been broken, they will not germinate well when planted. Dormancy of most species can be broken by storing the seeds at room temperature for several months before planting.

Origin

This label entry identifies the state or country in which the seed were grown. If the origin is unknown, the statement origin unknown must appear on the label.

Lot number

Lot number allows the seed producer and buyer to identify the specific unit from which the seed were taken. The term "lot" means a definite quantity of seed that is uniform throughout for the factors that appear on the label. This information is useful in cases of performance problems.

Net weight

This is the weight of the seeds only – minus the weight of the container.

Pure seed

This percentage tells the total weight of seeds of the kind and variety stated on the tag. If more than one kind or variety is named, the pure seed percentage of each component must be given.

Inert matter

Extraneous material, such as dirt, stems, leaves, and seed parts in the seed lot, is described in this percentage. Inert matter reduces the value of seed. It is best to choose seed with less than 2 percent inert matter.

Other crop seed

The percentage of the total weight from seed from a crop other than the kind and variety listed is indicated here. High-quality seed should contain no seed from other crops. The presence of other crop seeds in forage grasses can be particularly troublesome. Annual ryegrass in fescue is considered "other crop seed," not weed seed. The presence of 0.05 percent ryegrass in a forage translates into as many as 120 ryegrass seeds per pound of fescue. At a planting rate of 15 pounds per acre, this could mean as many as 1,800 ryegrass plants per acre of pasture.

Weed seed

Weed seed contamination is expressed as a percentage of the total weight. This classification includes seeds, bulblets, or tubers of plants recognised as common weeds by official regulations or by general agreement. Weed seeds are typically very small, so a small percentage of contaminants can translate into many weed plants. High-quality seed should contain less than 0.2 percent weed seed.

Seedsman or vendor

The name and address of the person or company labeling the seed are given on the seed label. They are responsible for the accuracy of the label.

Hard seed

Hard seed are viable, but have a seed coat that is impermeable to water. Where hard seeds are present, total germination percentages are customarily determined by combining germination and hard seed percentages.

Test date

The law requires that the germination test must have been made within nine months of the date the seed are sold or offered for sale (not counting the month of the test). The law also requires the individual owning the seed to keep the germination test date current, so carryover seed should be retested and re-labeled before the original test date expires.

Treated seed

Seeds that are treated with chemicals must be labeled to show the chemical used, and the words "Caution," "Poison," or "Poison Treated" must be written on the label, depending on the harmfulness of the seed treatment to humans. In addition, the statement, "Do not use for food, feed, or oil purposes," also must be printed on labels of treated seed, and the seed must be dyed a bright color.

Endophyte

The N.C. Seed Law does not require labeling for endophyte in tall fescue. However, if the seed are labeled, the N.C. Department of Agriculture and Consumer Services (NCDA&CS) does check to ensure levels of endophyte are correctly stated. If a discrepancy is found, a "stop sale" will be placed on the seed lot.

Noxious seed

The name and number of all noxious weed seed must be listed on the tag.

Storing Seed

High-quality forage seed can be stored successfully for more than one season if it is kept in a cool, dry environment. In general, the shelf life of grass seed stored at room temperature (77°F) is roughly two years. Orchardgrass and tall fescue, however, are exceptions and can be stored for only about one year. Legume seed storage potential is fairly good; most can maintain germination for two years, with the exception of crimson clover, which can be stored only one year.

Tips on storing

Avoid storing seeds in metal storage buildings or attics. Temperatures there can be extremely high during the summer, and it does not take long under these conditions for germination potential to be destroyed. Basements are well suited for seed storage as long as moisture can be controlled.

Forage seeds may be stored in freezers; however, freezing temperatures may kill Endophyte and Rhizobia. Seeds are a tempting food source for rodents and many insects. To protect seeds from rodents, store them in metal trash cans with the lid securely fixed. To control insects, tape no-pest strips on the inside of the lids. Do not insert the strip into the seed bag. Replace the strips every two to three months.

Vitality test

After storage, it is essential to test seed for germination at least one month before anticipated use. This will allow time to locate fresh seed if germination has declined to low levels during storage. The seed lab also can test seed lot viability using the tetrazolium chloride (TZ) test. TZ tests can be performed in 24 hours, giving growers an accurate and rapid evaluation of seed lot quality. There is a small charge for TZ tests.

Seeding rate

Seeding rates should be based on the amount of pure live seed (PLS) in the seed lot. This is particularly true for chaffy native species with fluffy seed coats or large awns or for seed that are difficult to clean. In the latter case, the seed contain large amounts of inert matter, which makes calculating seeding rates difficult. Suggesting seeding rates based on bulk seed weights can result in low plant populations. Common species like fescue, ryegrass, orchardgrass, and legumes generally have 95 percent or greater purity with germinations above 90 percent. Therefore, it is customary to suggest planting rates based on bulk seed (pounds of seed per acre).

PLS is calculated by multiplying seed lot germination by purity. For example, if germination is 90 percent and pure seed 90 percent, PLS is 81 percent [(90 x 90) ÷ 100]. The actual seeding rate based on PLS can be calculated by dividing the recommended rate by the PLS. If the recommended seeding rate is 15 pounds per acre, the actual seeding rate would be: 15 pounds per acre ¸ 0.81 = 18.5 pounds per acre.

At-Home Germination Test

Seed-testing facility perform an accurate evaluation of germination because a trained seed analyst can best judge germination potential. Many forage species also require special germination-test environments not readily available to growers. Some freshly harvested species require special pre-chill treatments before they will germinate. It is best to have a reliable evaluation of fresh seed performed by a qualified laboratory.

Growers who want to test carryover seed, not fresh seed, at home may use the following procedure:

1. Wet three paper towels with tap water. Hold the towels by one corner for several seconds to allow excess water to drain.

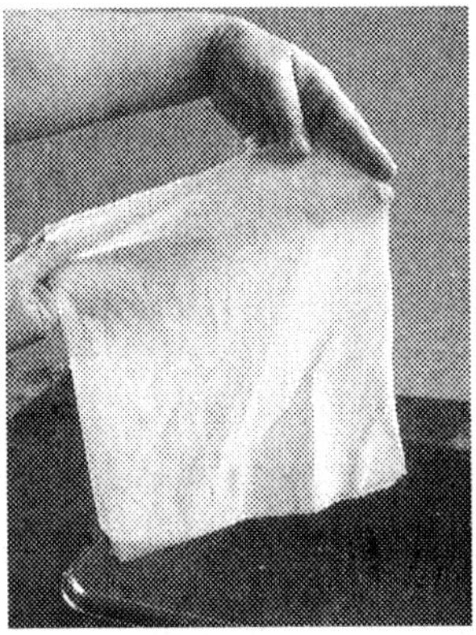

2. Place two towels together (back to back) and smooth them on a flat surface. Put the third towel to one side.

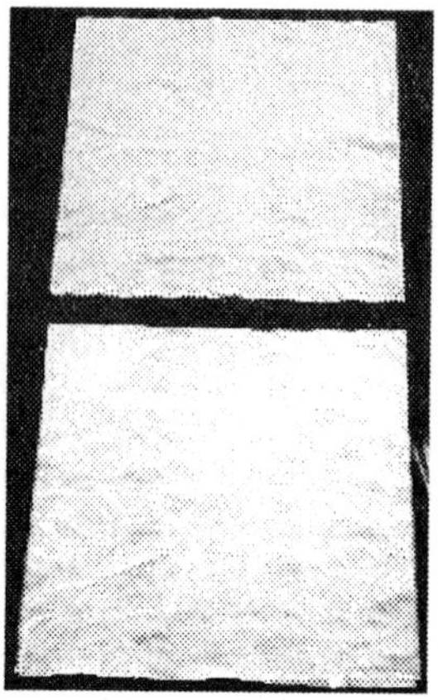

3. Place 100 seeds on top of the two flat, moist towels, and scatter the seeds evenly so that none is touching.

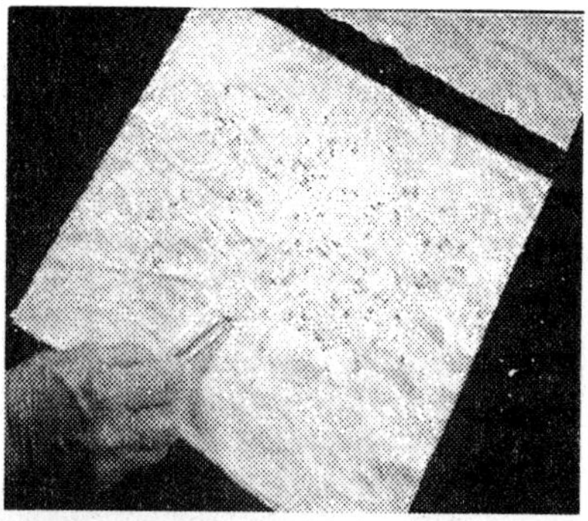

4. Cover the seeds and towels with the third towel.

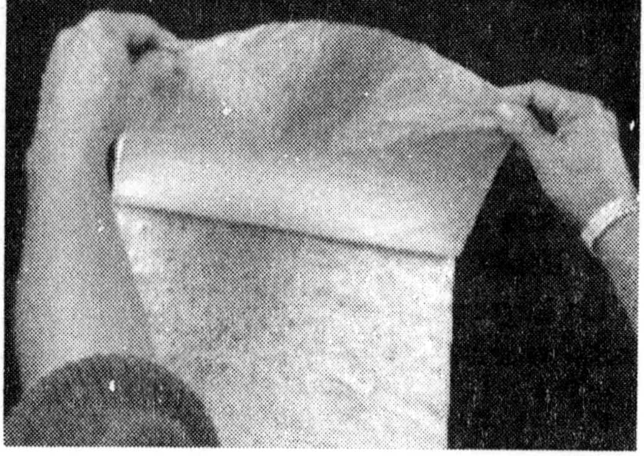

5. Gently roll the towels into a tube. The tube should not be too tight, for this will restrict seedling growth.

6. Place the tube in a large, clear plastic bag. Punch a few small holes in the bag to allow some air movement. Put the bag near a window. Arrange the bag so that the towel tubes are nearly vertical. (This allows the roots to grow downward and makes counting much easier). Do not place in direct sunlight. Too much sunlight might raise the temperature too high inside the plastic bag. Keep the towels moist, not saturated, throughout the germination period.

7. Take germination counts 7 to 14 days after planting. To be counted as germinated, the seedling should have a strong root and shoot.

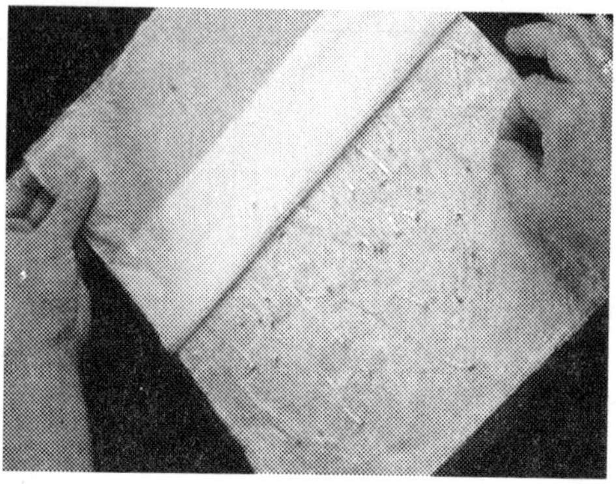

Clover and alfalfa may contain dormant hard seeds, which are typically counted as germinated in the test. To determine if seeds are dormant or dead after the allotted time in the germination towels, press the ungerminated seeds firmly into the towel with your finger. If the seed is hard, not mushy, it should be counted as germinated.

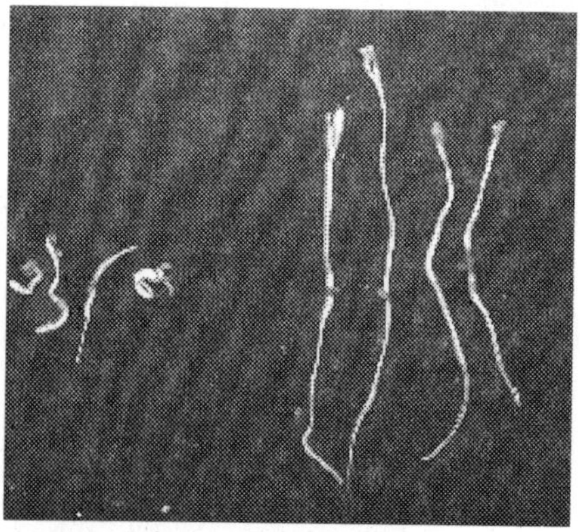

Alfalfa

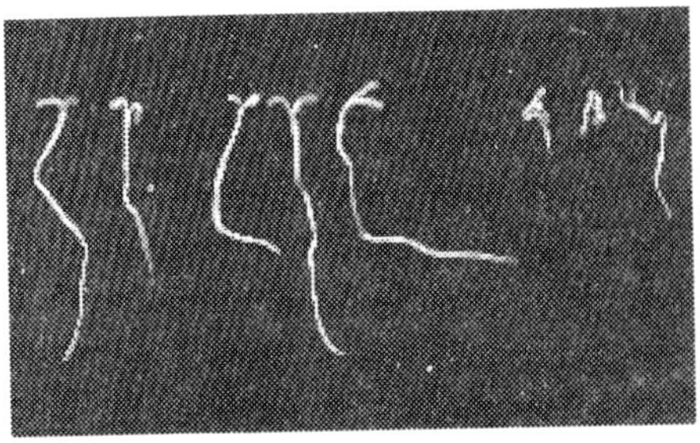

Clover

Germination, as determined by this at-home test, is the number of seeds out of the 100 tested that sprouted and had healthy seedlings (including the number of hard seeds).

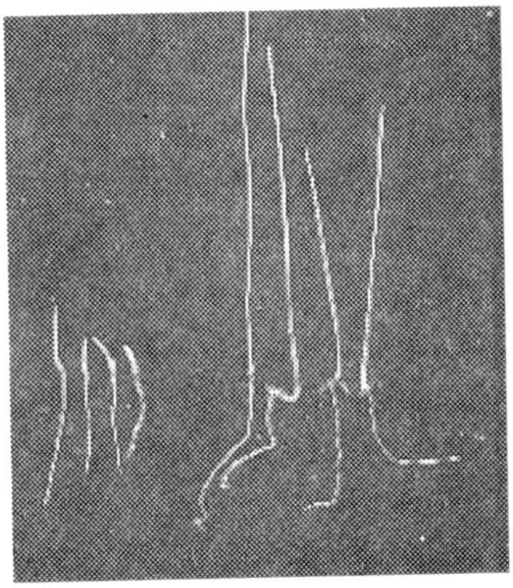

Fescue

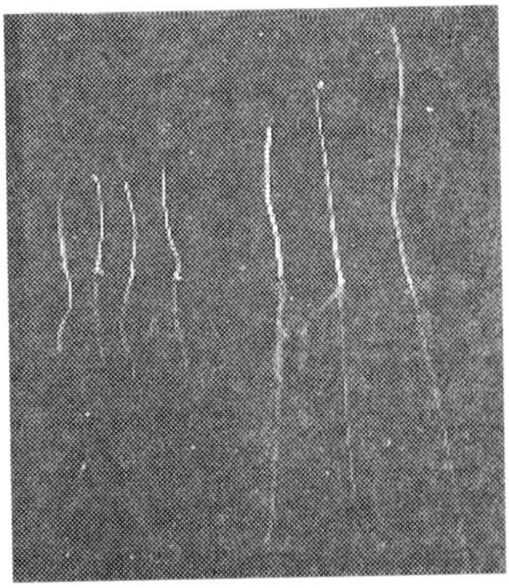

Ryegrass

Compare this percentage to the germination percentage on the original seed tag, and plan the seeding rate of the carryover seeds accordingly.

2

Seed Collection and Processing

Byproducts of the vast biological diversity and a broad range of climate, economic conditions, and political systems, tropical tree nurseries and seed-collecting organisations vary enormously in structure and in methods employed to collect and process seeds. These nurseries and organisations range from the most sophisticated and mechanised operations to tiny nurseries that use only hand labor and locally manufactured materials. All use valid and necessary means to solve the unique problems of each local situation. Seed collection and processing operations are molded by the objectives driving them.

Industrial objectives, such as supplying raw material to a productive mill or satisfying the demands of an export contract, require the production of abundant quantities of relatively high-quality seeds of one or a very small number of species. Usually great care is taken to select or produce seeds from highly productive genotypes. Most of these seed-collection organisations have generous budgets and a high degree of mechanisation. Seed suppliers that produce seed for sale both within their own countries and for export are required to produce the highest quality product possible. It must contain little debris and be free of insects and weed seeds.

Superior genotypes bring premium prices, but all lots must be forest-run (representing the average for wild trees) or collected from trees of selected phenotypes. There is constant pressure to offer as many species as possible, but costs tend to escalate with the number of species maintained in collections. These operations may employ one to many people, have varying degrees of mechanisation, and be government affiliated or private.

The most common objective of seed collection in the Tropics is to support local tree nurseries. These nurseries are the principal source of tree seedlings for homeowners, cities, farmers, and conservation and forest management organisations who use the seedlings for ornamental, agricultural, agroforestry, conservation, and silvicultural purposes. They typically maintain a relatively large inventory of the few species that are in continual demand (all of which are reliable seed producers) and a group of other species that changes as seed collection opportunities present themselves.

Seeds are usually collected by nursery employees or purchased from local seasonal collectors. Seeds are rarely stored for more than 1 year. Seed quality and handling are highly variable. Because wages are low in most of the Tropics and in some places the objective of providing employment supersedes that of producing seedlings, labor is often substituted for equipment or improvements requiring capital investment. Unfortunately, product quality and efficiency may decrease.

Collection and processing problems are species-driven. Hence, the greater the number of species handled, the greater the number of problems that must be solved. Unfortunately, some species are planted because their seed collection and management are easy, not because they are the best species available to meet the need. Many superb species are rarely or never planted because their seeds are difficult to collect or use. As experience in seed collection and management increases, many exciting species will be added to plantation inventories.

General Methods of Seed Collection

It is important to remember that fruit characteristics evolved to facilitate a seed dispersal strategy. Often a seed collection, extraction, or germination technique has simply mimicked the natural process.

From the Ground

The mainstay of seed collection is and will remain the picking up of the fruits or their seeds from the ground after they fall. This method is especially convenient for species with large or conspicuous fruits or seeds such as *Melia azedarach* L.,*Ormosia krugii* Urban,and *Terminalia catappa* L. Protective pods enable collection of fruits of such species as *Crescentia cajute* L., *Hymenaea courbaril* L., *Pterocarpus macrocarpus* Kurz, and *Senna spectabilis* (DC.) Irwin and Barneby, weeks to months after fruit-fall. Even the seeds or fruits of relatively small-seeded species, such as *Bucida buceras* L. and *Petitia domingensis* Jacq., can be collected from the ground if it is bare or paved, or if a tarp is placed under the tree just before fruit-fall. In those circumstances, vacuuming or sweeping can sometimes speed the collection process.

From the Tree

Another common means of collecting fruits and seeds is picking them from the trees. It is faster in many cases than collecting from the ground and it keeps the seeds cleaner. In addition, the seeds of many species are too small to pick up from the ground, and others are dispersed widely or consumed by animals and insects before they fall. Short trees may allow hand picking. Production can be accelerated by flailing into a basket attached to a picker's waist.

Picking must occur between physiological ripening (timed so the fruits will ripen in storage) and release by the tree or consumption by predators or dispersers. Ripeness is usually indicated by a color change of the fruit from some shade of green to an indicator color such as brown or red. In some cases, animal predation indicates ripeness. The sight of cockatoos feeding on the seeds indicates that *Agathis* seeds are ripe. The seeds of many species, such as *Albizia lebbeck* (Kunth) Harms and *Melia azedarach* L., remain in their fruits on the trees for weeks or months, greatly facilitating collection. Many of the eucalypts carry large quantities of fruits for extended periods that open quickly after fires. These fruits will also open after the clipped twigs dry. Species with serrotinous cones, such as *Pinus patula* Schiede & Deppe, produce cones that remain on the tree with viable seeds 1, 2, or more years after ripening.

Challenges

Tall Trees

Most species are too tall for hand picking. Pole pruners provide a convenient and inexpensive way to collect fruits from 2 m to about 9 m above the ground. Stepladders and easily portable, straight or extension ladders up to 7 m are also used. Fruits or seeds that shatter or detach easily, or seeds that are ejected from their fruits, can be collected by placing a tarp under the tree and flailing the tree with a long pole. In similar approaches, trees are shaken by hand (if small), by mechanical shaker, or by attaching a chain or rope to the trunk as high as convenient and to a vehicle or another tree, then jumping up and down on the chain or rope or pulling back and forth with a vehicle. A particular benefit of collecting seeds by shaking is that for species such as *Cordia alliodora* Oken, in which all the fruits or seeds do not mature at the same time, shaking releases the mature fruits while the immature fruits remain on the tree.

When short-statured species, such as *Acasia farnesiana* (L.) Willd., *Hibiscus tiliaceus* L., and *Moringa oleifera* Lam., and precociously flowering species such as *Spathodea campanulata* Beauv., invade cleared areas, they are short enough to pick by hand or pole pruner. Trees planted in open-grown situations sometimes grow low enough for easy harvest. Opengrown *Swietenia* spp. will usually bear many of its fruits within 9 m of the ground but does not do so in closed forests. Many species will continue to bear fruits after pruning to a low crown. Fruit trees, including *Mangifera indica* L. and *Citrus* spp., are routinely managed this way.

Crown-shaping trials with many tropical forest species should be conducted. The seeds of timber trees in natural forests are often borne near the tops of very tall trees. Traditionally, these seeds are collected from trees felled by logging operations. The method is excellent, if available. However, felling can shatter seed clusters and produce enough slash to hinder fruit retrieval. In some areas, trees have been destructively felled to obtain the seeds. Generally considered unacceptable ecologically and economically, this method may be beneficial if the seed trees are scheduled for removal to improve the stand.

The challenge of collecting seed from tall trees has inspired a number of approaches. Small quantities of seed, usually intended for breeding or provenance trials, have been collected by shooting down limbs with rifles or shotguns. Slingshot pellets and thrown sticks have been used to shatter small quantities of seed. Arrows have been used to draw lines over limbs to cut down or shake down small quantities of seed. Traditionally, local climbers were hired to ascend the trees and pick the desired seed in quantity with little or no safety equipment.

Any conscientious collector would now require the use of safety harnesses and belts, safety lines, and helmets. Treeclimbing spurs, tree bicycles, tethered sectional ladders, and other climbing aids can greatly accelerate an ascent up the tree bole. Once up the tree, a collector often uses a pruning pole to clip fruits on the ends of limbs or to saw off fruit-bearing limbs. Finally, seed collectors with generous budgets use highlift buckets on hard and nearly level ground to collect fruits from tall trees.

Widely Scattered Seed Trees

In tropical moist forests, seed trees of many species grow more than 1 km apart. Seed-producing adults are heavily exploited and rare species can often be very difficult to find. Random searches can be incredibly time consuming. Traditionally, seed collectors maintain a mental inventory of known seed trees or potential seed trees they have sighted during excursions through the forest over the years. Although it takes many years to accumulate this degree of experience, the system works well as long as these inventories are handed down to succeeding generations of collectors. The alternative is to maintain a written inventory of potential seed trees of at least the critical species along with maps of their locations. Modern technology can add efficiency and accuracy to this time-consuming process through the use of computer data bases and global positioning satellite (GPS) technology.

Genetic Impurity

A number of species hybridise freely with members of the same genus or varieties within the same species to produce undesirable or unpredictable offspring. The *Swietenias* have hybridised freely in Puerto Rico and the seeds of *Eucalyptus robusta* Sm., obtained from Brazil many years ago, introduced many hybrids that were generally inferior to the pure species. The timber tree *Hibiscus elatus* Sw. has been reported to hybridise with the shrubby *Hibiscus pernambucensis* Arruda in Jamaica. In the case of *Swietenia,* the hybrid seedlings can usually be separated from the parent species by leaf size. With *E. robusta,* sorting is impossible. The problem is solved by collecting from well-identified and isolated seed trees.

Unknown or Unpredictable Fruiting Seasons

In that portion of the Tropics with a strong wet-dry seasonal cycle, almost all species flower and fruit in certain well-defined seasons. Phenological studies can document these seasons and seed collection can be planned accordingly. However, in many species the flowering and fruiting dates may vary somewhat and the level of fruiting may vary tremendously from year to year, depending on seasonal rainfall amounts and patterns, and other factors such as wind or insect damage. Seed years are theoretically predictable from climatic conditions and can be successfully predicted from flowering and fruit set for a few species, such as *Pinus caribaea* sensu Small, non Morelet, which flowers one year and bears seeds the next.

In those portions of the Tropics with a relatively even distribution of rainfall, many species, such as *Roystonea* spp., *Ficus citrifolia* P. Miller, and *Hibiscus elatus* Sw., flower and fruit irregularly throughout the year. Some species, such as *Leucaena leucocephala* (Lam.) de Wit, bear seed more or less continuously in moist habitats, but seasonally in habitats with a strong wet-dry cycle. *Vochysia hondurensis* Sprague in Costa Rica bears fruit twice per year. Fruiting in some species is thoroughly unpredictable; individual trees fruit irregularly from year to year and by seasons and are not synchronised with others of their species. However, this can be an advantage.

If seeds are unavailable on one tree, they may be present on the next tree; or if seeds are unavailable in one locality, they may be available a few kilometers away. *Byrsonima spicata* (Cav.) Kunth, *Cordia sulcata* DC., and *Buchenavia tetraphylla* (Aubl.) Howard illustrate this behavior. Although *Swietenia macrophylla* G. King is usually seasonally synchronised even outside its native range, a few unsynchronised individuals produce some seeds during most of the year. Continuous flowering and fruiting and discontinuous/ irregular year-round fruiting is a natural strategy to avoid overloading the demand for pollinators and seed dispersers. Moreover, having a small percentage of the population out of synchrony helps avoid loss of regeneration to irregularities in normal rainfall patterns. To solve the problem of collecting from species with this diverse behavior, phenology should be recorded and collection activities should be planned by species. Collectors should not become discouraged by a few failures.

Delayed and Rare Fruiting

Many species, such as *Bertholletia excelsa* Humb. & Bonpl., do not bear fruit until they become large canopy dominants, a process that may take 50 years or more. In plantations, the process can be shortened to 15 to 25 years and grafted stock will bear fruits in as little as 6 years. Seed orchards exploit the tendency for open-grown trees to bear fruit more quickly and prolifically than forest trees. The abundance and ease of collection from open-grown trees has often led to excessive collection from phenotypically unproven trees in pastures and along streets. *Lagerstroemia speciosa* (L.) Pres. begin fruiting in as little as 3 years, but do not bear viable seeds until about 15 years. Most species of *Bambusa* flower and fruit in regional synchrony only once every few decades, and *Corypha umbraculifera* L., a palm, flowers once at the end of its long life and then dies.

The advantages to collecting seeds of species that can be stored during bumper crop years are numerous. The cost per unit of seed is lower; fewer seeds are damaged by insects; and the seed usually germinate at higher rates. A number of healthy exotic trees, such as *Araucaria heterophylla* (Salisb.) Franco in Puerto Rico, do not produce seeds in their new habitats, and seeds must be imported each season. Ringing, shallow girdling, stem strangulation, stem bending, root pruning, and water supply restriction has shown some promise in promoting seed production, although these methods ultimately injure the trees.

Animal Predation

Rodents, monkeys, birds, bats, and grazing animals can quickly eliminate a seed crop in a limited area. Parrots in Central America can consume an entire seed crop of *Acacia aneura* F. Muell. before it ripens. Although fences, screens, scarecrows, reflectors, and noise makers can successfully reduce or eliminate seed predation, they are usually practical only for seed orchards or concentrations of seed trees. Sometimes seeds can be harvested after they have become viable but before they or the fruits become attractive or accessible to animals. When forest species are scattered, collecting more widely and intensively appears to be the only way to obtain the needed seed stocks. In temperate areas, squirrel caches can be robbed; and in the Tropics, seeds are separated from the manure of predators that have been feeding on the fruit of the desired species.

Insect Infestation

Most species are attacked to some degree by seed insects. Occasionally, species such as *Prosopis juliflora* (Sw.) DC. and *Triplochiton scleroxylon* K. Schum. are so seriously attacked by insects that propagation is limited. In some cases, insecticides can be used to prevent attack and assure good seed crops. *Zanthoxylum flavum* Vahl seeds in Puerto Rico are reduced to <5 percent viability by a seed weevil. A conservation organisation was able to produce seeds free of insects and with good germination by spraying with insecticide (Rivera, personal communication). Many types of seeds should also be treated by fumigation, cold treatment, or insecticide application to eliminate insect damage during drying and storage.

Availability of Short Period

For various reasons, fruits or seeds of many species are ripe and available on trees for a very short time, often just a few days. The fruits of *Hyeronima oblonga* Muell. Arg. and *H. alchornioides* Allem. Diss. fall 3 to 4 days after maturity. In the final stage of ripening, *Pinus caribaea* sensu Small, non Morelet cones change color from green to brown, the cones open, and the seeds are dispersed quickly. Frequent field checks are essential to best time seed collection. Because individual trees of a species are often not closely synchronised, collecting from tree to tree can extend the collection season. Often, the collector can lengthen the period of seed collection by moving up an elevational gradient or across a moisture gradient.

Some species, such as *Maesopsis eminii* Engl. and *Pouteria* spp., picked up just before ripening, ripen in storage; thus the picking season is lengthened by a few days. The collector must know the species' traits, because some species, such as *Cordia alliodora* Oken, stop ripening as soon as they are detached from the tree. Collecting diptocarp seeds from the ground, for example, must be carefully timed because a delay of a few days can result in loss of viability.

Seed Processing

The basic method of removing seeds from fruit structures by hand is useful for research quantities of seed, and even highly productive for a few species such as *Swietenia* spp., but is too slow for most applications. Traditional methods such as flailing, trampling, grinding with a mortar and pestle, winnowing, floating, or screening are used to clean seeds of some species such as *Albizia procera* (Roxb.) Benth, *Casuarina* spp., *Eucalyptus* spp., *Leucaena leucocephala* (Lam.) deWit, and *Melaleuca* spp. Seeds in clusters and in brittle pods or capsules, such as those of *Acacia farnesiana* (L.) Willd., *Cassia javonica* L., *Parkinsonia aculeata* L., and *Zanthoxylum martinicense* (Lam.) DC., can often be cleaned by using standard research thrashing machines or by breaking them up by gentle hammermilling followed by screening and blowing. Usually, each stage of the operation is adjusted to accommodate the many seed sizes, shapes, and densities.

Challenges to Processing

Seed cleaning difficulities

The seeds of *Ochroma pyramidale* Cav. are covered with a silky wing ("silk") that is time consuming to remove by hand. In an innovative solution, the seeds were thinly spread over a large mesh screen and set on fire. As the silk burned, the seeds fell through and were scarified in the process. It is often adventitious to de-wing or eliminate other dry parts from seeds or fruits to reduce the bulk for storage or shipping. De-winging machines designed for *Pinus* spp. have also been used to de-wing other species and to remove some unwanted tissue, such as scale tissue and bract points in *Araucaria cunninghamii* Aiton ex D. Don. Because the seeds of *Swietenia* spp. are large and fragile, they are dewinged by hand, but other more resistant seeds, such as *Tectona grandis* L.f., may be cleaned by working sacks of seed against the ground with a foot, then removing debris by blowing.

Many species; such as *Dalbergia sissoo* Roxb., *Guazuma ulmifolia* Lam., *Prosopis juliflora* (Sw.) DC., Lam., and *Tectona grandis* L.f., have very tough pods or capsules. Some, such as *Pterocarpus macrocarpus* Kurz., have fragile seeds within. Others have fruit structures that tightly adhere to the seeds, as in some *Araucarias*. These seeds can sometimes be separated by hand using knives or clippers but not by any machines currently available. Usually, the pods or capsules are broken into small units and planted as seeds.

When seeds cannot be separated from fruits or debris, they can be planted within their fruits or mixed with debris. This method frequently requires thinning any resulting seedlings that appear in clusters. Sometimes, increased germination will justify hand cleaning small lots of seed, especially if the seeds are scarce or expensive or if labor is abundant. One ingenious method of cleaning the seeds of sweet podded legumes, such as *Acacia* spp., *Enterolobium cyclocarpum* (Jacq.) Griseb., and *Samanea saman* (Jacq.) Merril, involves feeding the pods to cattle or goats and extracting the cleaned (and scarified) seeds from the manure.

Another class of cleaning problems occurs when the seeds are encased in fleshy fruits. When the fruit is thin, seeds may sometimes be dried within the fruit and planted with the adhering fruit residues without affecting germination. In some species, adhering fruit can be removed or loosened by promoting rotting or allowing consumption by insects. If the fruits of *Acrocomia media* O.F. Cook are incubated under mulch for several months, the fleshy fruit is reduced to a husk and the seeds are not damaged. This method also afterripens the seeds. The same process, with a shorter duration, may be used with *Juglans jamaicensis* C. DC.

However, fruit must usually be removed soon after collection to facilitate storage and planting and to prevent damage to the seeds during putrification of the fruit. Numerous methods and machines for cleaning fleshy fruits are used. In many nurseries, most fleshy fruits are separated by hand, washed over screens, and dried. This method is the most rapid way to clean many large-seeded species such as *Mammea americana* L. and *Persea americana* Miller. Sometimes soaking the fruits in water for a period of time facilitates separating the seeds.

In wet sieving, the fruits are hand macerated against a screen just small enough to retain the seeds. The fruit pulp is continuously washed away by running water or by swirling the screen in standing water. The cleaned seed is then dried before storage. In a variation of this method, the fruits are pulped with a hand tool or machine before wet sieving. Flotation is also frequently used to separate fruit pulp, debris, and hollow seeds from filled seeds that sink.

Inconvenient seed sizes

The most familiar example of giant seed size is *Cocos nucifera* L. These giant seeds must remain in the husk, which limits human carrying to about 10 seeds and requires a great deal of storage and nursery space. Other species, such as *Caryocar villosum* (Aubl.) Pers. and *Mammea americana* L., have large seeds that also may be inconvenient to transport and store in large quantities. Despite these problems, the abundant stored reserves of large seeds often compensate for the inconveniences of size with their excellent germination and impressive early growth.

Very small seeds can also present problems. Because the seeds of *Eucalyptus* ssp. and *Melaleuca quinquenervia* (Cav.) Blake are so small, the fruits must be collected and transported before they open and disperse their seeds. Cleaning dry, small seeds is easy, but cleaning very fine seeds embedded in fleshy fruits, such as *Ficus* spp. and *Muntingia calabura* L, can be difficult. These seeds are usually cleaned by wet sieving and drying, sometimes followed by dry sieving or blowing.

Perishable seeds

Although usually thought of as a storage and seed handling problem, the short viability (as little as a few days) of some species must be considered during collection to facilitate rapid processing and planting or dispatch of seed to the intended user. Another class of seeds, known as recalcitrants, lose their ability to germinate upon drying. They are especially prevalent in humid forests and are more commonly associated with fleshy fruits. The process of germination often begins as soon as the fruit matures.

Whenever possible, seed operations with recalcitrant species, such as *Andira inermis* (W. Wright) Kunth ex DC., *Inga vera* Willd., *Persea americana* Miller, and *Thespesia grandiflora* DC., should harvest, clean, and plant the seeds on the same day or within a very short time. If these seeds are to be stored a short time at room temperature, they should be placed in a plastic bag or under wet sacks. Most of the species will withstand refrigeration and can be stored for one to several weeks. An alternative to seed storage is to germinate the seeds and maintain the seedlings in the nursery until needed. Although this is impractical with fast-growing species, slow-growing species such as *Guaiacum officinale* L. can be maintained an extra year in the nursery without difficulty.

Physical and chemical defenses

The fruits and seeds of a few species, such as *Pterocarpus angolensis* DC., are armed with spines or bristly hairs that make picking and extraction difficult. Using gloves and other protective clothing and possibly singeing or milling the fruits or seeds can facilitate seed handling. In some species, such as *Comocladia* ssp., *Hippomane mancinella* L., and *Sapium laurocerasus* Desf., all parts are toxic, at least when green and especially to sensitive individuals. Others, such as *Sterculia apetala* (Jacq.) Karst., have irritating or allergenic hairs. A large number of toxic and irritating species are scattered throughout the Tropics. Complete avoidance of these species by sensitive individuals and the use of disposable gloves, gauze masks, and other protective clothing by those who must work with them is strongly recommended.

3

Seed Storage

Most seed storage practices have developed empirically, often as a result of a combination of chance observation and pragmatic decisions on what can be achieved most easily by way of environmental control or modification. In agriculture, for example, the driving force for such empirical knowledge has been the simple practical requirement to store seeds reliably from harvest to the next production season. In silviculture, similarly, advice on storing tree seeds has long been available. For example, one and a half millenia ago it was recommended with regard to chestnut (probably *Castanea mollissima*) that "the fresh seeds should be stored in damp soil placed in the house, and during transport the seeds should be packed in a leather bag, because chestnut seeds will die when exposed to sun and wind".

Seed Storage Behaviour

Although far from complete, the world literature includes several comprehensive compendia which provide information on tree seed storage. Some compendia on seed storage behaviour were based on the duration of seed survival (whether air-dried or in soil or water) without considering the role of environmental factors on seed longevity. It was later found that certain species' seeds classified initially as "microbiotic" by Ewart or "short-lived" seeds can in fact be stored for long periods of time under more suitable conditions. Therefore, later compendia have classified species on the basis of seed desiccation tolerance and/or the response of seed survival or longevity to environment.

There are three different, discrete categories having been identified: orthodox seed storage behaviour; recalcitrant seed storage behaviour; and intermediate seed storage behaviour.

Orthodox Seed Storage Behaviour

Orthodox seeds can be dried without damage to low moisture contents and, over a wide range of environments, their longevity increases with decreases in seed storage moisture content and temperature in a quantifiable and predictable way. The latter is defined by the seed viability equation

$$v = K_i - p / 10^{K_E} - C_W \log_{10} m - C_H t - C_Q t^2, \qquad (1)$$

where v is probit percentage viability after p days in storage at m percent moisture content (w.b.), t°C, K_i is a constant specific to the seed lot, and K_E, C_W, C_H and C_Q are species viability constants. The constant Ki indicates the initial viability (in probits) of the seed accession when placed in storage. The value of K_i may vary considerably among different seed lots within a species, for example as result of differences in seed production environment, time of harvest, or genotype. The constants C_H and C_Q together describe the response of seed longevity to temperature.

The particular form of the relation between longevity and temperature is a negative curvilinear semi-logarithmic relation whereby the relative benefit to longevity of each 10 °C reduction in temperature declines, the cooler the temperature. A comparison of contrasting species, which included one tree species, found that the values of the temperature terms C_H and C_Q did not differ significantly among species. Similarly, the values of C_H and C_Q determined for four forest tree species (*Liquidambar styraciflua* L., *Pinus elliottii*, *Pinus taeda*, and *Platanus occidentalis*) by Bonner were 0.0306-0.0508 and 0.000328-0.000976, respectively, and thus close to the common estimates for diverse species of 0.0329 and 0.000478, respectively, determined previously. This suggests that seed longevity in all orthodox species may well show a similar quantitative response to storage temperature.

At one temperature, the gradient of the negative logarithmic relation between seed moisture content (m) and longevity provides the value of the constant C_W. Thus the value of this constant quantifies the effect of moisture content on longevity. The value of C_W differs considerably among species. For example, the values of C_W collated by Hong and others for 53 species from 24 families vary from 0.983 for *Pinus elliottii* to 6.305 for *Sorghum bicolor*. In general, species having seeds with a high starch content, e.g. cereals, have high values for C_W (around 6) but in oily seeds it is much lower, e.g. between 3.5 and 4 for onion (*Allium cepa*) and soyabean (*Glycine max*).

Since tree seeds are generally rich in oil in the embryo and/or endosperm, tree species tend to have lower values for C_W, e.g. 0.983 for *Pinus elliottii* and 4.23 for Norway maple (*Acer platanoides*). In one comparison of estimates of C_W, the mean value for forest tree species was 2.8 compared with a mean value of 4.72 for herbaceous species. One consequence is that to obtain the same relative increase in longevity it is necessary to dry orthodox tree and oily seeds more than starchy seeds from a given initial moisture content.

As a consequence of the negative logarithmic relation between seed moisture content and longevity, the relative benefit to longevity becomes greater for each successive reduction in moisture content. For example, based on the viability parameters estimated for *Ulmus carpinifolia* seeds by Tompsett, the effect of a reduction from 7 percent to 5 percent moisture content provides about a 2.7-fold increase in longevity whereas a reduction from 5 percent to 3 percent in seed moisture content increases seed longevity about 4.5-fold.

There appears, however, to be a common response of longevity to equilibrium relative humidity among contrasting crop species, at least, whereby longevity increases by a factor of about 2.2 for each 10 percent reduction in equilibrium relative humidity. However, we do not yet know whether this generalisation for seeds of arable and horticultural crops also applies to tree seeds.

There are two limits to the negative logarithmic relation between seed moisture content and seed longevity. One is an upper limit, above which seed longevity in hermetic storage is no longer reduced with further increases in moisture, and above which in aerated storage seed longevity increases with further increase in moisture content. The other is a lower limit, below which further reduction in moisture content no longer increases longevity in hermetic storage. The upper moisture-content limit appears to be 11 to 12 percent in pine (*Pinus elliottii* and *Pinus taeda*), 22 percent in elm (*Ulmus carpinifolia*) and *Araucaria columnaris*, and between 15 percent in lettuce (*Lactuca sativa*) to about 26 percent in durum wheat (*Triticum durum*) for crop species.

Despite wide variation among species in terms of moisture content, these values coincide with a water potential of about -14 MPa, i.e. the upper moisture-content limit to the viability equation occurs when seed moisture contents are in equilibrium with about 90 percent relative humidity (r.h.) at 20°C.

The lower limit varies substantially between crop species, e.g. between about 6 percent moisture content for pea (*Pisum sativum*) and mungbean (*Vigna radiata*), and about 2 percent for sunflower (*Helianthus annuus*). However, these variant moisture contents coincide with 10 to 12 percent equilibrium relative humidity at 20°C or with a seed water potential of about -350 MPa at this temperature.

In open storage (i.e. where seeds are exposed directly or indirectly to ambient relative humidity; e.g. in a paper bag, cloth bag, or thin, unsealed polythene bag which permits seed moisture content to be influenced by ambient relative humidity) at ambient temperature the longevity of orthodox seeds varies considerably. For example, viability of *Salix* spp. cannot be maintained beyond 10 days in open storage while seeds of *Cassia multijuga* were capable of 100 percent germination when tested after 158 years in open storage at room temperature in a Mediterranean climate.

For the maintenance of viability in long-term storage, *Salix* spp. require hermetic storage at low moisture contents and cool temperatures, e.g. sub-zero temperatures and moisture contents in equilibrium with 10 to 30 percent r.h. (relative humidity). Clearly, seeds of *Cassia multijuga* do not need such precise control of the storage environment in order to maintain viability in storage over the long term.

Provided estimates of the viability constants are available, the viability equation can be applied to determine suitable air-dry storage environments for a given duration of storage for a particular orthodox species. Since the same storage life can be achieved by different combinations of storage temperature and seed moisture content–the precise values depending upon species–there can be considerable flexibility.

Since the relative benefit to longevity from reducing temperature becomes less the lower the storage temperature, while the relative benefit to longevity from reducing moisture content becomes greater the lower the moisture content (within the wide range of environments over which the viability equation applies), it is often more cost-effective to reduce seed moisture content than to lower storage temperature.

This approach is of particular importance for seed storage centers where refrigeration to zero and sub-zero temperatures cannot be provided. In such situations it has been recommended that the seeds be dried to moisture contents in equilibrium with 10 to 12 percent r.h. at 20 °C and then stored hermetically at ambient or (preferably) cooler temperatures. This low-technology approach has been described by some as "ultra-dry" seed storage. Results from a 5-year investigation have shown that the longevity of ultra-dry seed at ambient temperatures is greater than that of crop seeds stored at 5 to 6 percent moisture content, at least for several crops. Similarly, over a much longer duration, 90 percent germination of seeds of barley (*Hordeum vulgare*) was reported when tested after 110 years in ultra-dry storage (3.1 percent moisture content) at 10 to 15°C. However, to date, ultra-dry seed storage has not been applied to tree seeds to our knowledge.

Successful storage of orthodox tree seeds over the long term under the FAO/IPGRI preferred conditions, i.e. 3 to 7 percent moisture content and -18 °C, has been realised for more than two decades in the Royal Botanic Gardens' Kew Seedbank and for about 7 years at the International Institute of Tropical Agriculture. For example, seeds of *Rhus verniciflua* and *Acacia tortilis* stored in the Kew Seedbank since 1970 and 1974, respectively, showed 100 percent germination in recent tests.

Long-term storage of orthodox tree seeds has also been achieved by cryopreservation (or cryostorage) at ultra-low temperatures from -80 °C to -196 °C with liquid nitrogen. For example, no loss in viability was reported for *Pinus ponderosa, Pseudotsuga menziesii, Thuja plicata,* and *Tsuga heterophylla* following 3 years of cryopreservation in liquid nitrogen. An essential first step in seed cryopreservation is the determination of optimum (safe) moisture contents for each orthodox species, particularly those with oily seeds. Wang and others collated information on the cryopreservation of orthodox tree seeds; it suggests that moisture contents between 3.8 and 11 percent are safe for at least short periods of time (from 4 days to 3 years).

However, a loss in viability of about 10 percent following 14 days storage in liquid nitrogen at 6 to 8 percent moisture content was reported for seeds of certain tree species. One possible cause of problems with liquid nitrogen seed storage is the requirement to optimise both the rate of cooling and the subsequent rate of warming; these must be determined empirically. Pretreatment with cryoprotectants sometimes improves seed survival in liquid nitrogen, but in some circumstances can be damaging. Mechanical injury to seed by direct immersion into liquid nitrogen causing loss in viability has also been reported for some species.

Finally, despite suggestions of infinite seed longevity by cryopreservation in liquid nitrogen, there is no firm evidence that reducing seed storage temperatures much below -20 °C results in further significant benefits to longevity, at least for tree seeds. For example, Tompsett found similar longevity for *Ulmus carpinifolia* seed stored at -13 °C and -75 °C.

For the maintenance of viability over the medium term, i.e. a period of 2 to 10 years or so, hermetic storage in a wide range of contrasting environments can be successful. For example, with relatively high moisture contents (in the air-dry range) combined with cool temperatures, e.g. 8 to 10 percent moisture content and 0 to 10 °C for most leguminous trees, or low moisture contents combined with relatively warm temperatures, e.g. 3 to 7 percent moisture content and 10 °C to ambient temperatures, e.g. Barton reported 33 percent germination for *Ulmus americana* L. seeds following 4 years of hermetic storage at 3 percent moisture content at room temperature. Ultra-dry seed storage may be applicable to tree seeds for medium-term storage at ambient temperatures. For example, Joseph and Brinkman recommended that seeds of *Betula* spp. be stored at moisture contents as low as 0.6 to 1 percent at room temperature.

In addition to air-dry storage over a very wide range of different environments, successful storage of orthodox seeds over the short-term, from a few months to 1 or 2 years, can also be achieved by moist storage at cool temperatures. This approach is commonly used for tree species that exhibit strong dormancy such as those adapted to temperate climates (i.e. temperate latitudes and high altitudes of the Tropics) since prechilling during storage can remove dormancy. For example, several orthodox tree species of temperate climates, such as *Carya* spp., *Chionanthus virginicus, Euonymus* spp., *Hamamelis virginiana, Ilex* spp., *Juglans* spp., *Juniperus* spp., *Liriodendron tulipifera, Nyssa* spp., *Picea* spp., and *Taxus* spp. can be stored moist at 3 to 5 °C.

Similarly, orthodox seeds of several species that are native to high altitudes in the Tropics and show considerable dormancy can also be stored moist at 3 to 5 °C. For example, seeds of *Prunus africana,* native to high mountains between 900 m and 3400 m in Africa, and *Michelia champaca* L. and *Michelia compressa,* native to high mountains of Asia, can be stored moist at 4 °C for 1 year. In addition, seeds of some pioneer forest trees that have an absolute requirement for light in order to germinate can also be stored moist in the absence of light at ambient temperature for long periods of time. For example, seeds of *Piper aequale* can be maintained for 7 years fully imbibed in darkness without loss in viability.

Recalcitrant Seed Storage Bahaviour

Recalcitrant seeds cannot be dried without damage and so their longevity does not respond to air-dry environment in the manner described by the seed viability equation (1). When freshly harvested recalcitrant seeds are dried, viability is at first slightly reduced as moisture is lost, but then begins to be reduced considerably at a certain moisture content, termed the "critical moisture content" or "lowest safe moisture content". If drying continues further, viability is eventually reduced to zero. Critical moisture contents for loss of viability on desiccation vary greatly among the recalcitrant species, among cultivars and seed lots, and depending upon the stage of seed maturity at time of collection.

The critical moisture content may also vary with the method of seed drying. The values of the "lowest safe moisture content" vary between extremes of about 23 percent for cocoa (*Theobroma cacao*) to 61.5 percent for *Avicennia marina*. Despite this variation, these moisture contents are equivalent to a relatively narrow band of relative humidities of 96 to 98 percent, or seed water potentials of about -1.5 MPa to -5 MPa.

There is no satisfactory method of maintaining the viability of recalcitrant seeds over the medium and long term. Since they cannot be dried neither can they be stored at subzero temperatures, because they would be then killed by freezing injury resulting from ice formation. The longevity of recalcitrant seeds is short, from a few weeks to a few months for species adapted to tropical environments, and up to about 3 years for several species adapted to temperate environments. However, if the optimum storage environments are carefully determined, longevity of several tropical recalcitrant seeds can be extended to 3 years, at least for species of *Symphonia globulifera*.

The principle of successful moist seed storage for recalcitrant seeds is that seeds must be maintained at moisture contents close to that at which they are shed, with continuous access to oxygen; these circumstances minimise seed deterioration since repair mechanisms can operate. Under these conditions (high seed moisture and available oxygen), however, seeds tend to germinate. Clearly it is also essential that the conditions should prevent or at least delay germination. It is therefore easier to store recalcitrant species with dormant seeds (either primary dormancy or induced dormancy) than with nondormant seeds under such conditions.

For nondormant seeds, as shown by most tropical tree seeds at maturity, low temperatures can reduce the rates of both seed deterioration and germination provided they remain above the value that results in chilling damage or the lower value at which ice crystallisation occurs. For example, chilling injury occurs at between 5 and 10 °C for *Shorea roxburghii*, 5 and 12 °C for *Symphonia globulifera* and *Hopea odorata* Roxb. Determination of the optimum temperature for imbibed seed storage is required for each recalcitrant species. A suggested protocol for determining these values has been described elsewhere.

The optimum temperature appears to be that at which nondormant seeds remain alive but are unable to germinate, i.e. the base temperature for germination. For example, the optimum storage temperatures determined in this way are 10 °C for *Shorea roxburghii*, 12 °C for *Mangifera indica*, and 15 °C for *Hopea odorata* and *Symphonia globulifera*. However, recalcitrant seeds of species adapted to temperate climates show considerable dormancy which requires long periods of prechilling at 2 to 5°C to overcome. Furthermore, seeds of many temperate species with recalcitrant seed storage behaviour have the ability to germinate at 2 to 5 °C, and thus moist storage of such recalcitrant seeds at 0 to 10 °C may result in germination during storage.

In such cases, germination may be prevented by reducing seed moisture content slightly (by about 5 percent below that of fresh seeds), or either reducing the storage temperature to below the optimum prechilling temperature (e.g. 0 to -3 °C) or increasing it to the base temperature for germination of dormant seeds. For example, the viability of seeds of sycamore (*Acer pseudoplatanus*) and *Quercus robur* can be maintained for 3 years at 24 to 32 percent and 40 to 45 percent moisture content, respectively, at -1 to -3 °C, and horsechestnut (*Aesculus hippocastanum*) for 3 years if moist dormant seeds are stored at 16 °C, although in this case subsequent prechilling is required to remove dormancy.

In practical terms, species with recalcitrant seeds can therefore be divided into two subcategories: those adapted to tropical climates, and those adapted to temperate climates (i.e. temperate latitudes, or high altitudes in the Tropics). In general, the viability of recalcitrant seeds can be maintained (albeit only for limited periods in aerated conditions) at moisture contents just less than fully imbibed, i.e. about 2 to 5 percent below those of fresh seeds, or in equilibrium with 98 to 99 percent r.h., at optimum storage temperatures which vary from about 7 °C to 17 °C among species adapted to tropical climates, and between about -3 °C and 5 °C among many of those adapted to temperate climates.

Maintaining recalcitrant seeds fully imbibed or subimbibed, with continuous aeration, and at the same time preventing germination and fungal contamination, is difficult. Aeration can result in loss of seed moisture, and respiration can quickly deplete oxygen. The storage medium is therefore very important for recalcitrant seeds. It should fulfill two functions: first, to maintain seed moisture constant at high values; second, to allow diffusion of sufficient oxygen to the moist seeds. The storage of moist recalcitrant seeds in damp charcoal, sawdust, or moist sand is generally reported to be more efficient than storage in polyethlene bags.

Storage at or near the harvest moisture con-tent in media such as sawdust (at 16 percent moisture content) or perlite (at 0 to 4 percent moisture content) within suitable containers, such as open-weave sacks or bags, placed in a highhumidity room has been recommended.

Coating recalcitrant seeds with alginate or alginate with ABA was also reported to extend seed lifespan by a factor of up to 3 or 4.

Cryopreservation of zygotic embryos has been reported to be successful for a number of tree species that show orthodox, intermediate, or recalcitrant seed storage behaviour. For successful cryopreservation, excised embryos or embryonic axes must survive desiccation below the threshold freezable moisture content of about 18 to 33 percent, below which value there is no freezable water for ice formation in ultra-low temperatures. Embryos and embryonic axes of some recalcitrant species are able to survive desiccation to lower moisture contents than are the whole seeds.

For example, fresh seeds (36 percent moisture content) of *Hevea brasiliensis* tolerated desiccation to 20 percent moisture content, but no seeds survived further desiccation to 15 percent moisture content. However, after desiccation to 16 percent moisture content followed by 16 hours cryopreservation in liquid nitrogen, 87 percent and 69 percent, respectively, of the excised embryonic axes (55 percent moisture content) survived. Unfortunately, embryos of many recalcitrant seeds are damaged by desiccation before seed moisture content is reduced below those values at which ice crystallisation no longer occurs, and so the cryopreservation of excised embryos of these species results in death, either by ice formation for embryos at moisture contents above the threshold freezable moisture content, or by desiccation to moisture contents below that threshold.

Fast drying (also termed "flash" drying) may allow excised embryos to survive desiccation to lower moisture contents than those dried more slowly within intact seeds. Excised embryos or embryonic axes can be dried quickly in a laminar airflow cabinet at room temperature or under a flow of compressed air at about 9 liters min^{-1}. After flash drying of embryonic axes extracted from mature seeds of *Landolphia kirkii* reduced moisture content from 67 percent to 23 percent in 30 minutes, 90 percent of the excised embryonic axes were able to germinate. Fast drying with silica gel or with an aseptic air current has allowed excised embryonic axes to survive desiccation to a lower value than that achieved by vacuum drying.

The stage of maturity of the seeds from which the embryos are excised is an important factor influencing survival following desiccation: embryos excised from mature seeds not only survived desiccation to lower moisture contents than immature embryos, but also survived subsequent cryopreservation in liquid nitrogen. Fast freezing can allow desiccated embryos to better survive cryopreservation in liquid nitrogen. In contrast, slow freezing with immature excised embryos of cocoa has been reported to be more successful than fast freezing. The specimen to be frozen must be as small as possible, and pretreatment of embryos with cryoprotectants, such as DMSO, proline, sucrose, and glycerol, are important factors influencing the survival of excised embryos (or embryonic axes) in liquid nitrogen.

Results reported for *Aesculus* spp., *Araucaria hunsteinii, Artocarpus heterophyllus* Lam., *Castanea sativa, Citrus hystrix, Coffea liberica, Cocos nucifera, Dimocarpus longan, Hevea brasiliensis, Landolphia kirkii, Quercus* spp., *Trichilia dregeana* and *Theobroma cacao* have shown the feasibility of the cryopreservation of excised embryos or embryonic axes of recalcitrant seeds. However, others have reported a lack of success using the same species. For example, experiments with excised embryos of *Artocarpus heterophyllus, Landolphia kirkii,* and *Quercus robur* found that none survived cryopreservation in liquid nitrogen.

Intermediate Seed Storage Behaviour

Although the term "intermediate seed storage behaviour" was introduced only at the beginning of this decade, this third category of seed storage behaviour was hinted at more than eight decades ago. Elliott divided temperate forest tree seeds into three classes:

i) those that can be dried,

ii) those that can survive partial drying, and

iii) those that can rarely be dried at all.

The definition of intermediate seed storage behaviour is based on the response of longevity to storage environment. In seeds showing intermediate seed storage behaviour, the trend toward increasing longevity the lower the seed storage moisture content (within the air-dry range) is reversed at a comparatively high moisture content, whereby at lower moisture contents longevity is reduced. But in such species it is also often (but not always) observed that seeds may be damaged immediately by desiccation to relatively low moisture contents, about 7 to 12 percent moisture content depending on species.

The critical moisture contents of intermediate seeds below which more rapid loss in viability occurs during hermetic storage varies considerably with species, degree of maturity, and method of seed extraction and/or postcollection handling. In general, seeds extracted from fruits at maturity tolerate desiccation to moisture contents in equilibrium with about 40 to 50 percent r.h., i.e., about 10 percent moisture content for arabica coffee (*Coffea arabica*) and 7 percent moisture content for *Citrus* spp.

A further feature of intermediate seeds of tropical origin is the fact that the longevity of dry seeds (7 to 10 percent moisture content) is reduced with reduction in storage temperature below about 10 °C. In such cases, then, there is an optimum air-dry storage environment for the maintenance of seed viability. In arabica coffee this is about 10 °C with 10 to 11 percent moisture content. It is now known that seeds of neem (*Azadirachta indica* A. Juss), *Cinnamomum subavenium*, several *Citrus* species, *Coffea arabica*, robusta coffee (*Coffea canephora*), oil palm (*Elaeis guineensis*) Khaya senegalensis, Lindera megaphylla, Neolitsea parvigemma, and Swietenia *macrophylla* show intermediate seed storage behaviour.

Among about 7,000 species examined in a recent survey, about 134 species are believed to show intermediate seed storage behaviour. Many of these species are trees of tropical origin, but a few are herbaceous. For example, that survey suggests that the following important tree species may show intermediate seed storage behaviour: Acer macrophyllum, Afrocarpus gracilior, Agathis macrophylla, Araucaria columnaris, A. nemorosa, A. rulei, A. scopulorum, Bertholettia excelsa, Bixa orellana, Chrysophyllum cainito, Dacrycarpus dacrydioides, Dipterocarpus alatus Roxb. & G. Don, D. intricatus, D. tuberculatus, Dipteryx odorata, Rhapidophyllum hystrix, and Virola surinamensis.

From the point of view of optimum air-dry seed storage environments, it may be helpful to distinguish between species with intermediate seed storage behaviour adapted to tropical environments and those adapted to temperate environments. For example, intermediate seeds of tropical origin such as arabica coffee and papaya (*Carica papaya*) can be stored at moisture contents in equilibrium with 50 percent r.h. (9 to 10 percent moisture content) and 10 °C for up to 5 and 6 years, respectively, without loss in viability.

The viability of intermediate seeds of temperate origin is also maintained well at moisture contents in equilibrium with about 50 percent r.h., but at cooler temperatures of 5 °C to -10 °C. For example, seeds of *Corylus avellana* can be stored hermetically at -5 °C with 10 to 13 percent moisture content. Species showing intermediate seed storage behaviour can be stored successfully over the medium term, therefore, provided optimum environments have been defined and can be maintained.

Like orthodox seeds, intermediate seeds can also be stored moist at cool temperatures if germination is either prevented or delayed. For example, seeds of arabica coffee and tea (*Camellia sinensis*) were stored moist at 15 °C and 1 °C, respectively, for 2.5 and 6 years, respectively, with little loss in viability. The situation regarding the cryopreservation of intermediate seeds is a little confused. Despite reports of the immediate death of (whole) seeds of species with intermediate seed storage behaviour following cryopreservation in liquid nitrogen, e.g. arabica coffee, *Corylus avellana*, *Corylus cornuta*, oil palm, and *Roystonea regia* (Kunth) O.F. Cook, there have been several reports of the survival of dry seeds of species with intermediate seed storage behaviour following immersion in liquid nitrogen, e.g. *Camellia sinensis*, *Carica papaya*, *Citrus aurantifolia*, and *Citrus halimii*. This implies that it may be possible to improve cryopreservation techniques to make them suitable for intermediate seeds.

Since the whole seeds of species with intermediate seed storage behaviour tolerate desiccation to relatively low moisture content (7 to 10 percent), desiccated embryos may have a greater chance to survive cryopreservation in liquid nitrogen than is the case for recalcitrant seeds. Reports of successful cryostorage of excised embryos of *Azadirachta indica, Camellia sinensis, Citrus aurantifolia* and *Citrus halimii, Coffea arabica, Corylus avellana, Elaeis guineensis, Howea forsteriana, Poncirus trifoliata,* and *Veitchia merrilli* have shown the potential for long-term seed storage under such conditions.

Desiccation Tolerance and Potential Seed Longevity

Application of the seed viability equation (1) reveals that initial seed quality (K_i)–potential longevity–plays an important role in ensuring good seed survival in storage. The higher the value of K_i the longer the period of viability for a given storage environment. During seed development and maturation, the value of the seed lot constant K_I–so the potential longevity of orthodox seeds–increases markedly.

There is now good evidence in a wide range of contrasting orthodox crop species that potential longevity continues to increase during the period after the end of seed filling, as seeds mature. There is also good evidence that the ability of orthodox seeds to tolerate desiccation to very low moisture contents also improves considerably during the late stages of seed maturation, and indeed one study has, therefore, shown a strong association in developing and maturing seeds between potential longevity (K_i) and the ability of seeds to tolerate desiccation to very low moisture contents.

The ability of seeds to tolerate enforced rapid desiccation has been reported to increase during seed development and maturation of tree seeds with orthodox, intermediate, and recalcitrant seed storage behaviour. Orthodox seeds of several crop, wild herb, and tree species are not able to tolerate rapid enforced desiccation to low moisture contents (4 to 5 percent) until some time after mass maturity once maturation drying has reduced seed moisture content on the mother plant substantially.

Maximum tolerance of rapid enforced desiccation occurs some time after mass maturity, probably at natural dehiscence. In addition, the slow desiccation of immature seeds harvested before mass maturity, and well before maturation drying occurs naturally, enables tolerance of rapid desiccation to be acquired. For example, in Norway maple (*Acer platanoides*), rapid drying (by reducing moisture content from 57.3 percent to 9 percent within 1 day, and to 5 percent within 4 days) of the seeds harvested at mass maturity reduced viability from 100 percent to 61 percent and 38 percent at 8.1 percent and 4.7 percent moisture content, respectively, while for seeds that were first dried slowly for 32 days, during which period seed moisture content was reduced from 57.3 percent to 29.9 percent, subsequent rapid drying to 3.5 percent moisture content (within 4 days) resulted in 93 percent viability. The slow drying of seeds or fruits harvested before or at mass maturity has also been reported to increase seed vigor and potential longevity in several crop and wild species.

In addition, desiccation tolerance also decreases when seeds are initially subjected to conditions under which germination is initiated. For example, prechilling, moist storage, presoaking, fermentation treatment for seed extraction, and seed priming can reduce desiccation tolerance, and hence alter seed storage behaviour. Similarly, seeds that are produced in unfavourable environments may show reduced desiccation tolerance to low moisture contents. For example, seeds of a japonica rice (*Oryza sativa* subsp. *japonica*) produced in a hot environment of 32/24°C (day/night) showed consistently poorer desiccation tolerance at each stage of maturation drying than those produced in a cooler environment of 28/20 °C.

Prediction of Seed Storage Behaviour

Methods, and indeed feasibility, of seed storage depend upon the seed storage behaviour shown by a particular species. A simple, two-stage protocol to determine seed storage behaviour has been suggested elsewhere. This task is, however, considerable given that there are about 250,000 species of flowering plants. Estimation of likely seed storage behaviour would therefore be helpful, and several approaches to doing this have been developed. No single criterion can provide a satisfactory estimation of likely seed storage behaviour, but combining information for at least four of the six factors described below may be worthwhile.

Plant Ecology and Seed Storage Behaviour

There appears to be an association between plant ecology and seed storage behaviour. From information on seed storage behaviour collated for almost 7,000 species from 251 families, it is evident that species that show recalcitrant seed storage behaviour do not occur naturally in (i.e. originate from) arid habitats, that is desert and savanna. In such environments, the majority of plant species show orthodox seed storage behaviour, while a few may show intermediate seed storage behaviour. It is clear, however, that further generalisations are not possible. In particular, for example, at the other ecological extreme it is quite clear that not all species native to moist habitats, rain forests, flooded forests, or aquatic environments show recalcitrant seed storage behaviour; all three categories of seed storage behaviour can be found among the species native to such moist environments.

Taxonomic Classification

Orthodox seed storage behaviour can be found in all species within a few families (e.g. Chenopodiaceae, Combretaceae, Compositae, Labiatae, Solanaceae, Pinaceae), or almost all species in others with only a few exceptional species (e.g. Leguminosae, Graminae, Cucurbitaceae, Cruciferae and Rosaceae). However, most families of flowering plants including tree species have species with two (e.g. Anacardiaceae, Dipterocarpaceae, Euphorbiaceae, etc.) or three different categories of seed storage behaviour (e.g. Meliaceae). Furthermore, seed storage behaviour can differ among species within a genus.

Plant, Fruit, or Seed Characteristcs

Certain generalisations appear possible concerning associations between fruit and seed characteristics and seed storage behaviour. For example, orthodox seed storage behaviour is shown by species that produce achenes, many-seeded berries, many-seeded dehiscent capsules, many dry-seeded pods (but not arillate), many dry-seeded follicles, schizocarps, and utricles. Most species that produce siliques (one exception) and caryopses (three known exceptions) also produce orthodox seeds. On the other hand, however, all three categories of seed storage behaviour can be found among species that produce one seed or a few large seeds (from 1 to 10 seeds) per fruit, or many arillate seeds per fruit, among drupes, pods, capsules, berries, and nuts.

Seed size

Recalcitrant seeds often tend to be larger than intermediate seeds, which in turn tend (on average) to be larger than orthodox seeds. However, there is a very wide range of seed weights or sizes over which seeds of different species show all three types of seed storage behaviour.

Seed moisture content at shedding

A recent survey shows that seed moisture contents at maturity or shedding for species with recalcitrant, intermediate, and orthodox seed storage behaviour overlap considerably. No generalisation on seed storage behaviour is possible for species in which seeds at maturity or shedding have between about 25 percent and 55 percent moisture content.

Seed shape

There appears to be a weak association between seed shape and seed storage behaviour. For example, recalcitrant seeds are often spherical or oval. Seeds that are thin and flat, shapes that aid natural maturation drying, tend to show orthodox seed storage behaviour. Flat seeds tend to show orthodox seed storage behaviour in *Acer* spp. , but show both orthodox and intermediate seed storage behaviour in Meliaceae. The spherical form is present in all three categories of seed storage behaviour.

All of the single criterion suggested above are not sufficiently reliable to be very useful as indicators of likely seed storage behaviour. However, it may be possible to develop a multiple-criteria predictive framework in order to suggest likely seed storage behaviour. For example, combining information on the four criteria–seed weight, seed moisture content at shedding, seed shape, and taxonomy can form the basis of a guide to likely seed storage behaviour among *Acer* spp. Similarly, the four criteria of seed weight, seed moisture content at maturity, seed shape, and plant ecology may provide a guide to the estimation of seed storage behaviour in Meliaceae.

For example, from the two positions designated A for seed weight (TSW = 4300 to 7700 g), ecology (moist evergreen forest of India), seed shape (convex, fleshy aril) of *Amoora wallichii*–information from Luna –we suggest that this species is likely to show recalcitrant seed storage behaviour. Similarly, from the two positions designated B for seed weight, seed shape (convex, fleshy aril), and ecology (rain forest of East Africa) of *Turraeanthus africanus*, we suggest that recalcitrant seed storage behaviour is likely in this species also.

Furthermore, the information available on natural habitat of the two species can also be used to suggest likely suitable moist seed storage conditions. Given that the *Amoora wallichii* tree is described as "frost-hardy", and that *Turraeanthus africanus* grows at an altitude of 1525 m, we suggest that chilling injury is unlikely if the seeds of both species are stored moist at about 10°C.

4

Collection and Storage of Tree Seeds

It is rather extraordinary to think that one small seed can grow into a tree that will live for hundreds of years, reaching a size several metres in diameter. But seeds are delicate, as all farmers know. To maintain their potential to sprout, (or "viability"), seed must be collected, stored, and treated properly. Knowledge of a few simple steps can make you a skilled tree seed collector. Following these steps will ensure that your seeds germinate well, and that you will get the kinds of high quality trees you are seeking from your efforts and at no expense.

1. How to select good mother trees.
2. Suitable time of seed collection.
3. Proper seed collection techniques.
4. Careful seed handling and extraction.
5. Good packaging and storage.
6. Effective pre-sowing treatment methods.

How to Select Good Mother Trees

Prior to seed collection, you will need to select and mark good mother trees. These are the trees you will use as sources of high quality seed. Here are the major characteristics that determine a good mother tree:

1. Healthy and free of diseases and insects
2. Nearly mature
3. Good producers of the desired product
4. Growing in the midst of a healthy stand of the same species.

To choose the right mother trees, remember this simple rule:

— A tree's off-spring will usually resemble its mother.

Therefore:

— If you want straight trunks, choose a straight-trunked mother tree.

— For a multi-trunk fodder tree, select multi-trunk mother trees.

— For trees which tolerate drought, or flooding, select good looking mother trees from dry or flooded sites, etc..

For timber trees:

— Mother trees should be fast growing, very straight, and have few and thin branches (Figure 1).

For fodder, good mother trees should:

— Be fast growing.

— Have many branches.

— Have several trunks.

— Grow new leaves quickly after fodder harvesting.

Trees which have leaves all year round produce more fodder than those which lose their leaves part of the year.

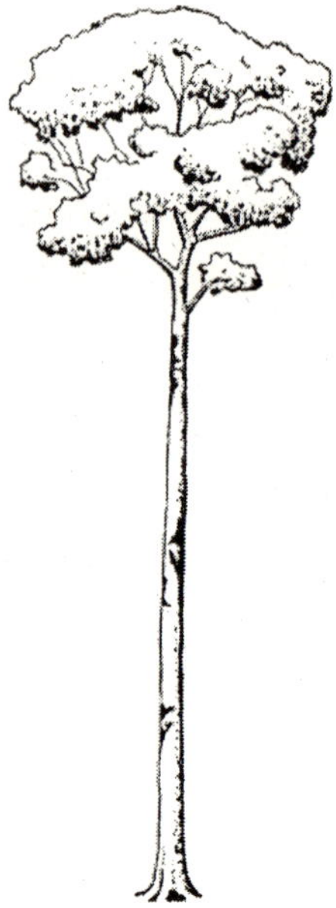

Figure 1. A good mother tree for timber.

The most important factors in selecting mother trees for fodder are:

— Fast production of leaf matter and pods preferred by local animals.

— Ability of the tree to recover after cutting (Figure 2).

For fruit trees:

— Collect seed from trees of local varieties producing good quantities of tasty, healthy fruit of marketable size.

— Low branching trees may be preferable as mother trees. It is easy to pick fruits from low branches.

Figure 2. A good mother tree for fodder.

For fruit trees, high quality varieties are usually grafted onto native, or hardy rootstocks.

Figure 3. A good mother tree for fruits.

To get a high quality fruit tree, these steps should be done:

1. Grow seedlings of drought tolerant and disease resistant mother trees.
2. Then graft a branch of a high quality variety onto the seedling when it is strong and well-established.
3. Keep your eyes open for trees of high quality varieties near your home. From these, you can obtain scions to use for grafting. Otherwise, obtain scions from commercial fruit nurseries (Figure 3).

Timing of Seed Collection

The optimum time for seed collection is as soon as the seed is mature. For most types of trees, seed pods turn brown when the seed inside is mature. Pods begin falling to the ground when seed is ripe.

Some types, however, must be collected when pods are still on the branches. Some, like *Grevillea robusta,* have light seed bearing fruits which will open on the tree and allow seed to blow away. In other cases, seed will over-ripen and not germinate well or be susceptible to insect attack if fruits or pods stay on the tree too long. Neem and *Acacia* sp. are examples of this. In these cases, collection must be done when pods or fruits are still on the tree.

Seed of some species, like *Combretum* and *Terminalia,* appear brown even when unripe. Careful observation must be done to ensure that the seed inside the pods is mature before harvesting. Mature seed is usually brown. If the seed coat is opened when ripe, the seed inside will usually be light in color, and only slightly moist. Most seed pods will crack open when seeds are mature. Your ability to know when a seed is perfectly ripe will develop through observation and experience.

How to Collect Seeds

After selecting and marking good mother trees, several seed collection methods can be used.

Collecting from Natural Seed Fall

This is the simplest way to collect seed. It does not require skilled labour. Collection from natural seed fall is suitable for trees with large fruits, pods, and seeds e.g. *Tectona, Gmelina* and some *Dipterocarps*.

The following tools will be helpful:

— Rake

— Sieve

— Seed container

— Large canvas, cloth or plastic sheet

Follow these steps:

1. Clear the ground beneath the tree of leaves, branches, and weeds before seeds begin to fall. This will make seed collection easier.

   Or,

   spread plastic sheets, cloth or canvas under the mother trees so that the seeds will fall onto them.

2. Use a rake to gather the seeds and collect them daily.

   Or,

   fold sheets to collect seeds daily. Chances of insect attack and fungal infection which could occur if seeds are left on the ground too long will be minimized.

2. Trample with feet, toss around, or beat pods inside a sack to speed up seed separation. Pods which do not easily split open may be opened manually, or threshed by pestling.
3. Clean seeds by blowing or winnowing.
4. Immerse seeds shortly in a container of water.
5. Discard those that float. They are not viable.
6. Drain the water, and re-dry the seeds. When seeds are to be planted immediately, re-drying is unnecessary.

Fleshy Fruits

1. If possible, seed extraction should be done at home.
2. Do not delay extraction for too long and never store fruit piled up and undisturbed. (Heating and fermentation in the pile could kill the seeds or reduce their quality).
3. Immerse in water until the fleshy tissue becomes soft (1-2 days).
4. Scrape, crush, or rub seeds with hands to separate throughly from pulpy flesh. (Be careful not to be so rough as to damage the seed).
5. Remove all floating seeds and pulp.
6. Drain off the water.
7. Re-wash the seed.
8. Air dry for 2 days.
9. Re-clean the seeds by winnowing.

Seed Drying

If seeds will be stored for future use, they must be sun dried.

1. Spread the seeds thinly and evenly on a mat, canvas sheet, light-colored plastic sheet, (a black sheet may cause seed to overheat), winnowing basket, or screen where the sun shines all day (Figure 5).

Figure 5. Drying seeds on a plastic sheet

2. Stir and turn the seeds 4-5 times a day for uniform drying. If possible keep seeds (especially moist ones) shaded during intense heat. (noon to 2:00 PM.)
3. Before it rains or gets dark, take the seeds indoors.
4. Drying will take 1-3 days, depending on how wet the seeds are.
5. Protect the seeds from rodents and birds during drying.

The same drying procedure may be used for seeds of most fruit trees. Exceptions are seeds of rambutan, durian, mangosteen, mango, jackfruit, avocado, rubber, cacao and mahogany. These should be air dried only for a day or two before storing, or they must be planted immediately after extraction.

Seed Storage

Under ordinary room conditions (= open storage), viability of many seeds is reduced by half within 6 months.

— Seeds with harder seed coats tend to live longer than those with thin coats.

— For improved storability, seed moisture and storage temperature must be kept low and controlled.

Follow this Simple Procedure Step-by-step:

1. Store only new, mature, healthy and well-dried seeds.
2. Keep them in dry and cool place to extend their viability.
3. Seeds easily re-absorb moisture. To maintain dryness, keep seeds in air-tight containers like tin cans or glass jars with tight fitting lids.
4. Put in some moisture absorbing material. Dry wood ash, dry charcoal, powdered milk, toasted (cooled) rice, or small pieces of newspaper are all good. The drying material should take up about one-fourth of the container space.
5. Label the containers with the type of seed, place, and date of collection.
6. If possible, include the initial percent viability of the seeds. To do this, plant some seed to see how many germinate. If 8 of 10 germinate, for example, percent viability is 80%. This information will help you to learn about how much each type of seed loses viability between collection and planting.

7. Protect seeds from insects and fungi. Before storing in containers, mix with dry ash, powdered seeds of black pepper or neem leaves. Or use extract of neem, peanut, castor bean, or cotton: 1 teaspoon oil/1 kg seed. Or use naphthalene balls: 1 or 2 pieces/10 kg seed.
8. Protect from rodents and birds during storage.

Use of Sealed Containers

— After seed has been dried properly, store it in tin cans, metal boxes, glass jars, or plastic bags or container with lids that can be sealed (Figure 6).

Points to remember when using Sealed containers:

— Never seal in moist seed.

— Use airtight containers.

— Make sure that the container is clean.

— Open the seed containers only when necessary.

Some species have seeds which do not tolerate drying or cold temperatures. Examples of these are rambutan, durian, mangosteen, mango, jackfruit, avocado, rubber, cacao, mahogany, and neem. They can only be stored from several days, up to a few months under ordinary room conditions. It is best, therefore, to plant these kinds soon after collection. But if they must be stored:

1. Do not dry the seeds completely. Leave them a little moist. Aim for a dry seed coat, but a moist seed kernel.
2. Fill plastic bags half and half with seed, and slightly damp charcoal, peat moss, sawdust, or sand.

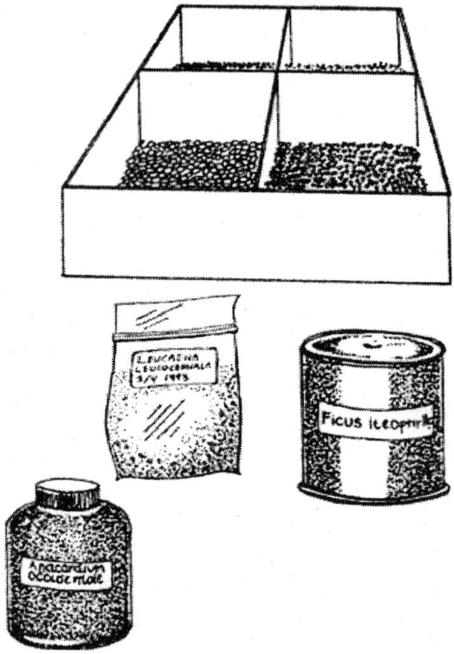

Figure 6. Seed storage containers: metal box, tincan, plastic or glass jar.

3. Keep the bags in a cool place.
4. Open the bags for one-half hour each day to permit air to enter.

Seed Treatments to Hasten Germination

Many tree seeds have hard seed coats which are impermeable to water and air, or which prevent the emergence of seed parts. Simple treatment is required to:

— Hasten seed germination

— Ensure rapid and uniform seedling establishment

— Shorten the time of exposure of potted seeds to pests and other stresses.

Ways of Seed Treatment

Tap Water Treatment

1. Soak seeds in tap water for 12-48 hours (depending on how hard the seed coat is) before planting.

Figure 7. Tap water treatment

Figure 8. Hot water treatment

2. If seeds do not absorb water, another treatment must be used (Seeds which have absorbed water will swell.) (Figure 7).

Hot Water Treatment

1. Pour hot water (boiled, and then cooled for about 5-10 minutes to 80 °C) into a container with seeds. (Use 10 parts of water to 1 part of seed.)
2. Let stand for 3-10 minutes or until water cools off.
3. If seeds have not swelled, leave to soak overnight (Figure 8).

Mechanical Scarification

1. Nick off seed coat with a knife or nail clipper, or

rub with sand paper, file or any rough material, taking care not to injure the internal portion, especially the radical (root end).

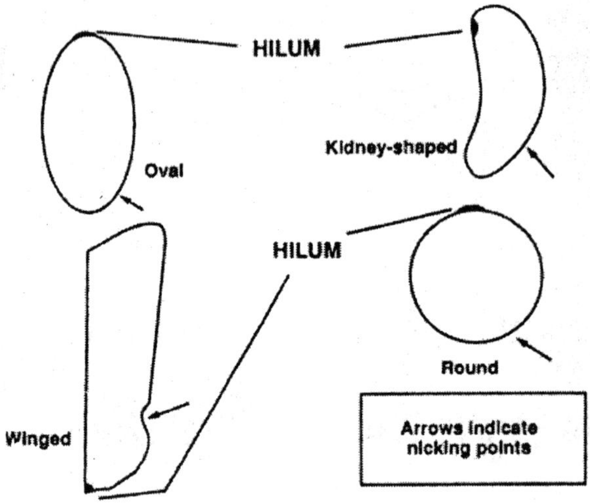

Figure 9. Scarification points by seed shape.

2. This technique is practical for small seed lots (Figure 9).

5

Seed Treatment

Seed Treatments have played and are still playing a significant role in the history of mankind and the ability to stave off hunger and starvation in improving the establishment of healthy crops leading to better yields.

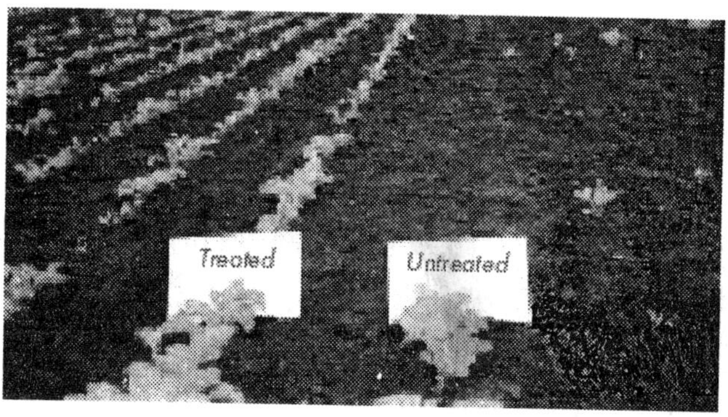

Beets with a seed treatment on the left and untreated on the right

Seed Treatments are the use and application of biological, physical and chemical agents and techniques used with seed that provide seed and plant protection and improve the establishment of healthy crops.

First Seed Treatments

Some of the first recorded uses of Seed Treatment are

from the Egyptian and Roman periods and involved a technique of using sap from onion (*Allium spp.*). In the middle ages, Seed Treatments were made with liquid manure and chlorine salts. Salt water treatments have been used since the mid-1600s and the first copper products were introduced in the mid-1700s. The "still used today" technology of hot water treatment is documented from as early as 1765 in Wittenberg, Germany. Seed was placed in 45°C water for 2 hours, this provided control of certain fungal pathogens on the outer surface of the seed.

Modern Seed Treatments

The seed industry is continuously in a time of transition and improvement. Key milestones in the history of Modern Seed Treatment are the introduction and ban of arsenic (used from 1740 until 1808), the introduction and ban of Mercury (used from 1915 until 1982).

Approx. 2000 B.C. to 100 A.C.	First soaking techniques: use of sap of onion or cypress (Egypt, Greece, Roman Empire)
Middle Ages	Soaking in chlorine salts and manure
1600's	Soaking in salt water
Mid-1700's	Introduction of copper salts
1740	Introduction of arsenic
1765	Soaking in hot water (Germany)
1808	Ban of arsenic
1915	Introduction of organo-mercurics
1960's	Introduction of first systemic fungicide
1970's	First systemic fungicide against air-borne pathogens
1982	Ban of organo-mercurics in Western Europe
1990's	Introduction of new modern fungicides and insecticides

The launch of the first systemic compound in 1960; until then, Seed Treatments had been only seed sterilants and did not move into the plant. During the 1970's the first systemic Seed Treatment fungicides for air-borne pathogens were introduced. In the 1990's, new modern classes of fungicide and insecticide chemistry were launched.

The Seed Industry has a long history of not only treating seeds but also of waste management of treated seeds. As early as 1786, there is documentation of a ban against feeding and grinding treated seed, understandable, because arsenic was a popular Seed Treatment in this era.

Today, the waste management of non-usable treated seeds, empty containers, and waste water are priorities for Seed Treaters and Seed Producers. The Seed and Seed Treatment Industries have a long history of partnership and dedication in bringing to the farmer high quality seed. The seed is not only as pest and disease free as possible, but treated to provide the possibility of pest and disease control, when needed, during germination and emergence of the young plant and early growth of the crop.

The objectives of modern Seed Treatment products are the superior control of certain insects and diseases while improving crop safety leading to good establishment of healthy and vigorous plants. Modern Seed Treatment formulations must also contribute to improvements in farmers' and workers' safety and stewardship of the environment.

Today's modern Seed Treatment products have to meet high safety and efficacy standards. The newest active substances and formulations provide long lasting, broad spectrum, systemic control of diseases and insects (depending on the specific active ingredient). Modern formulated Seed Treatment products used by the Farmer, Seed Treater, and Seed Producer are often sophisticated with several active ingredients and special wetting agents and colorants that are safe to the seed, the environment and the user.

Modern Seed Treatment products are able to deliver high levels of efficacy for the control of early season insects or diseases at a much reduced usage rate compared to many foliar or soil applied alternatives. Seed Treatment usage with today's compounds reduces the land surface exposed to the active ingredient. Imagine the comparison of 1 hectare of land with a foliar or soil applied application of active ingredient—results in 10,000 m^2 of land in contact with the active ingredient. If the application was used in-furrow, then the exposed land surface could be reduced to $500m^2$. But, the use of Seed Treatment would leave only 58m2 of surface exposed to the active ingredient, minimising environmental impact. Additionally, seed treatments have less risk of impact on non-target organisms and drift.

Integrated Pest Management

The use of Seed Treatment is consistent with the concept of Integrated Pest Management. The farmer must strive for the best pest management method that is safe, environmentally sound, and scientifically proven. Without the use of Seed Treatment, farmers would have great difficulty to control certain seed-borne and early season seedling diseases and insects and would have to resort to more expensive and less environmentally acceptable methods.

The FAO International Code of Conduct on the Distribution and Use of Pesticides (Article 2), states,

"Integrated Pest Management (IPM) means a pest management system that, in the context of the associated environment and the population dynamics of the pest species, utilises all suitable techniques and methods in as compatible a manner as possible and maintains the pest populations at levels below those causing economically unacceptable damage or loss." Seed Treatment can be used as a sound basis for a successful Integrated Pest Management Program.

Application of Seed Treatment

Seed Treatment is a term that describes both products and processes. The usage of specific products and specific techniques can improve the growth environment for the seed, seedling, and young plant. Seed Treatment complexity ranges from a basic dressing, to coatings and pelleting:

Seed Dressing: The most common method of Seed Treatment. The seed is dressed with either a dry formulation or wet treated with a slurry or liquid formulation. Dressings are applied both on-farm and industrially

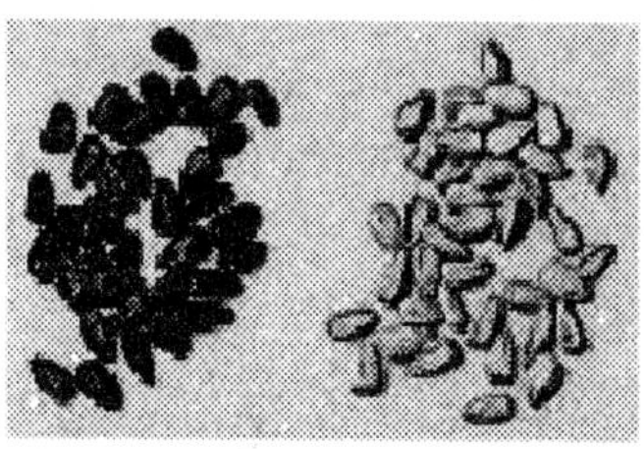

Seed Coating: A special binder is used with a formulation to enhance adherence to the seed. Coatings require advanced treatment technology.

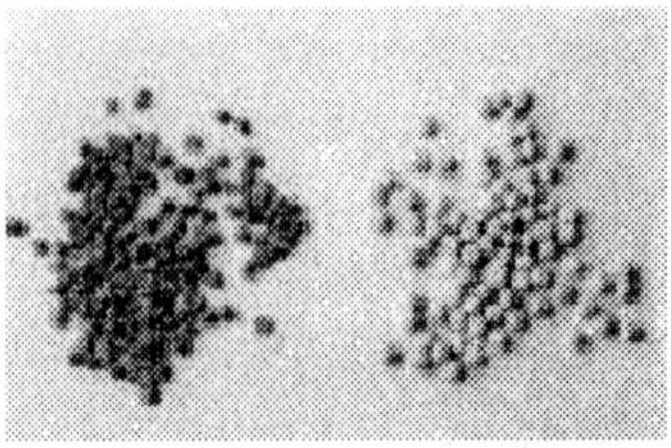

Seed Pelleting: The most sophisticated Seed Treatment technology, resulting in changing the physical shape of a seed to enhance plantability and handling. Pelleting requires specialised application machinery and techniques and is the most expensive of the applications.

Depending on the type of desired treatment, various treating machines can be used.

A simple machine, only for a basic dressing, and for small quantities is the auger type machine used in cereals.

The more complex the treatment, the more cost is involved. Many modern treatment machines can provide specific formulation and dosing by way of a direct injection process. The most sophisticated treatment machines can provide the various formulations and additives as layers on the seed, which allows precise placement of the active ingredient.

Farmers, distributors, officials, and regulatory authorities have high standards for the Seed Industry regarding the quality of a Seed Treatment and its application to the seed. The expectations are that Seed

Treatment will provide high performance and safety in processes and products.

Industrial treaters are available for high throughput and varying levels of customisation. They can be continuous flow or batch type and have the possibility to apply dressings, pelletings and coatings.

The Seed Industry recognises that Seed Treatment supports "Added Seed Value" (e.g., safe emergence, seedling establishment, crop stands, yield, and crop quality). Seed Treatments complement and protect genetic improvements, function as part of IPM (Integrated Pest Management), are safe and economical in use, and reduce personal and environmental hazards. Seed Treatments, compared to conventional crop protection products, offer competitive costs, reduced application efforts and save time. Seed Treatment is sometimes the only timing and delivery route for the control of certain seed-borne and early season seedling diseases or pests, which without the use of Seed Treatment, would be very difficult to control.

The Seed Industry and Seed Treatment Applicators make an effort to reach an optimum application effect and consistency, as supported within a "Total Quality Management" (TQM) system. A TQM approach defines the treatment process, handling of treated seed, worker protective measures, protective measures for the environment, application techniques, operational techniques and information requirements to facilitate the safe and legal treatment of seed, handling of treated seed, and use and disposal of treated seed.

The TQM approach should also provide transparent seed quality and performance information that is addressed to users. It is generally understood that Seed Treatment requires adequate quality standards and necessary application methods. These should be agreed between the involved parties case-by-case and accepted as a balance between potential high standards and justified costs to achieve the standards.

6

Seed Quality for Seedling Production

Seed quality has great impact on the quality of planting stock. For the last 20 years, the technology of producing seedlings has advanced greatly. Parallel to this advancement, seed quality also has improved dramatically.

Cone Collection

Careful attention to cone collection is critical to obtaining good quality forest-tree seed. Successful collection depends on understanding seed maturation and dispersal characteristics of each species, knowing local weather trends, and evaluat—ing crop quality, harvesting procedures, and cone-storage methods.

Seed Maturation

Cone collection should be gin only when seed is mature. Immature seed can bring about various problems including (1) slow and incomplete germination, (2) low-vigor seed, resulting in smaller seedlings, (3) greater susceptibility to disease, (4) reduced storage capability, and (5) increased incidence of abnormal seedlings. In addition, extraction of immature seed is more difficult than that of mature seed. Various maturation indicators, reflecting visual, physical, biochemical, or climatic changes, can be used effectively to prevent harvest of immature seed.

Cone color, bract color, seed wing color, scale color, and color and firmness of embryo and megagametophyte can be visual indicators of seed maturity. These indicators, though indirect and subjective, have proved reasonably practical in many instances. Cone moisture content, cone specific gravity, and embryo development can be physical indicators of seed maturity. Loss of cone and seed moisture is closely associated with seed ripening, and the decrease in cone moisture content and cone specific gravity has been used to indicate maturity. Of these two indicators, specific gravity (SG) is usually preferred because it can easily be determined in the field.

This method has been successfully applied to various pines (Pinus spp.) and true firs (Abies spp.) using flotation liquids such as water (SG = 1.0) and various mixtures of kerosene (SG = 0.80), light motor oil (SG = 0.88), and linseed oil (SG = 0.93). The ratio of embryo length to embryo cavity length, which can be determined quickly in the field with a sharp knife and a 10X magnifying hand lens, also can be used to judge maturity.

Changes occurring within conifer seeds can be biochemical indicators of seed maturity. On the basis of observed correlation of reducing sugar content and germination, Rediske recommended that Douglas-fir cone collection be initiated when reducing sugar content has fallen to 13 mg/g of seed weight. In a subsequent study, Rediske and Nicholson found that, in noble fir, the increase in crude fat content is more closely related to seed maturation and recommended the threshold value of 250 mg/g of seed for beginning cone collection. Although measuring biochemical indicators is time consuming and requires special laboratory equipment, it is thought to be more reliable than methods based on visual observation.

Changes in temperature, particularly during the summer in which seeds mature, can strongly influence the rate of seed maturation and are used as climatic indicators. Consequently, degree-day summations should be potentially more reliable than calendar date, especially at high latitude or high altitude, where summer temperature may limit seed development. Tanaka and Cameron reported that 1,310 degreedays are required for ponderosa pine seed to mature at high elevations in southeastern Oregon. Zasada related cone and seed development in white spruce [*Picea glauca* (Moench) *Voss*] to summer heat-sum and found that 625 degree- days were required to produce cones that could be successfully after-ripened in Alaska.

Maturation indicators

In US, maturation indicators for the six major coniferous species in the Northwest are summarized in Table 1. Those used by the one nursery and six seed-processing plants involved in cone collection are, in order of frequency: cone, wing, and scale color, firmness of embryo and megagametophyte, and embryo development. Somewhat surprisingly, seed moisture content and specific gravity are not currently used, probably indicating that visual observation of the above characteristics is preferred because it is less time consuming.

Table 1. Cone and seed maturation indicators for the six major conifers in the Northwest

Species	Maturation indicators
Douglas-fir	Reducing sugar 13 mg/g or less
	Embryo:cavity length ratio greater than 90%
	Browning of cone bracts
	Cone moisture content lower than 50%
	Firm, nonmilky megagametophyte enclosing a yellowish-green embryo
	Main harvest period of squirrels
Ponderosa pine	Specific gravity 0.85 or less (central Idaho)
	Specific gravity 0.84 or less (California)
	Specific gravity 0.94 to 0.99 (South Dakota)
	Specific gravity 0.88 or less (Arizona and New Mexico)
	Heat -sum 1,000 to 1,110 degree-days
Lodgepole pine	Specific gravity 0.43 to 0.89
Noble fir	Specific gravity 0.90 or less
	Crude fat 0.2 5 g/g of seed
White fir	Specific gravity 0.96 or less, uniformly brown seed wing, embryos pale yellow-green, 94% of the embryos fully elongated
Western hemlock	Brown cones with red-brown tips
	Cones opening after drying

One seed plant extensively relies on biochemical indicators, using crude fat for noble fir and ponderosa pine and reducing sugar for Douglas-fir; on the basis of past experience, these biochemical indicators seem highly reliable.

Seed Dispersal

Although seed-maturation characteristics have been exten—sively studied, little is known about timing of seed dispersal in relation to cone characteristics or climatic variables. Most observations relate the timing of seed dispersal to calendar dates, but such correlations may be of little value in the field because of yearly variation in weather patterns. Together with knowledge of seed maturation rate, approximate seed-dispersal dates could be of practical importance to cone collectors. The field observations made by our laboratory also indicated that the earlier seed matured, the more quickly it started disseminating, probably due to faster drying of cones. Similar observations should be of value in capturing the maximum seed yield of other conifers that have a responsive reflex of cone scales.

Artificial Ripening

It is important that cone collection be initiated after seed has attained full maturity. However, immature seed can be artificially ripened during cone storage in certain species. Artificial ripening has been successful on noble fir, grand fir [*Abies grandis* (Doug]. *ex D. Don*) *Lindl.]*, white fir, and Nordmann fir [*Abies nordmanniana* (Stev.) *Spach*]. Because of this potential increase in germination during cone storage, true fir cones are usually stored longer than those of other conifers before seed extraction. Douglas-fir, several species of pines, and white spruce have also shown increased germination during artificial ripening. However, despite the findings of various researchers and the potential benefits, artificial ripening has not been extensively used for conifers other than true firs in the Northwest-probably because there is more risk of poorer germination and reduced seed yield in other species.

Weather

Weather conditions significantly impact cone collection. Except for pines with serotinous cones or some cypresses (*Cupressus spp.*) or junipers (*Juniperus spp.*) for which year-round collection is possible, the optimum cone-collection period for most conifers at any given location is relatively short. This period occurs sometime between late summer and late fall but could vary by up to 2 to 3 weeks depending on weather conditions. For example, if the snow melts late at high elevations during a cool spring, flowering may be so late that seed maturation could be delayed significantly. A hot, dry summer may shorten the optimum cone-collection period by causing early seed fall, whereas cool, rainy conditions may delay it. Seed generally ripens earlier at lower elevations and on south and west slopes and later at higher elevations and on north and east slopes.

In addition to general spring and summer weather trends that determine seed-maturation and dispersal patterns, weather conditions during the cone-collection period itself are also important for the cone-harvesting operation. High winds or rain may preclude tree climbing, disrupt access to collection areas, and reduce pickers' productivity. In many areas in the Northwest, a drying east wind during the fall collection period may cause seed to disseminate too quickly, thereby reducing seed yield. For these reasons, daily forecasts and 5-day out -looks are valuable aids to coordinating cone-collection activities.

Crop Quality

Once seed maturity has been determined, the quality of cone crops to be harvested must be evaluated. This generally is done by estimating the number of good seeds present in several representative cones, sliced lengthwise with a sharp knife. Cones of Douglas-fir, western hemlock, and pines are sliced through the center; those of true firs are cut lengthwise ¼ to ½ inch to one side of center. A variety, of knife assemblies is available for slicing conifer cones. Minimum acceptable seed-count requirements may vary from year to year according to supply and demand. Average good seed counts are 6 for Douglas-fir, 8 for western hemlock, 10 for ponderosa and lodgepole pine, and more than 50% of the seed (if seed has good appearance) for noble and white fir.

Lodgepole pine in certain areas produces cones that are very hard and, therefore, difficult to section. To extract seeds, such cones can be dipped in boiling water for 10 seconds, then placed in an oven at 65°C for 3 to 4 hours. A minim um of 20 filled seeds per cone is required before a crop can be harvested. In addition to the filled-seed count, damage by biotic agents such as insects and disease, climatic extremes, or other abnormalities also should be assessed because these affect seed yield and are important factors in selecting areas from which to collect. Dobbs et al. do not recommend collection if more than 50% of seeds are damaged. Several articles may be of help in identifying and assessing insect [59, 74] and disease damage.

Collecting Methods

Cones are collected from western conifers: (1) by climbing standing trees, (2) from felled trees, and (3) from squirrel caches. Collecting cones from standing trees-the surest method to harvest seed of known origin, qua lity, and maturity-is often time consuming, expensive, and dangerous. Cones can be picked much more easily from felled trees in logged areas, but pickers should ascertain whether seeds were sufficiently mature when the trees were felled.

Cones should be picked immediately after felling so as to minimize seed loss due to cone opening or mammal, bird, and insect damage. Squirrelcached cones are easy to collect, but their use is sometimes questioned because the source and quality of the crop tree are not known. No evidence suggests, however, that seeds collected by squirrels are inferior to those collected by other means. Helicopter collection has been experimentally tested in Canada by Dobbs et al.. Mechanical tree shakers, regularly used on southern pines, have been tried only experimentally for western conifers. Although not easily adaptable to Northwest terrains for natural-stand collection, mechanized cone collection should play an important future role when western seed orchards are in full production.

Cone Storage

Cones are stored (1) because processing equipment is not usually capable of extracting seeds from all harvested cones at once ; (2) to decrease cone moisture content, thereby reducing kiln drying time; and (3) to artificially ripen seeds of species such as true firs and improve seed-germination potential. In large-scale cone collection, cones are usually placed in burlap bags, which are stored either temporarily near collection sites or in storage sheds at the extractory.

However, great care should be exercised to maintain seed quality during cone storage. Burlap bags should not be filled to the tops, so that cone scales can fully expand upon drying; if scales cannot open sufficiently, seed extraction may be severely impaired. Burlap bags should not be stacked up in large piles; this can lead to seed losses due to overheating or to insect and disease damage. Warm, moist environments can harm seed quality ; hence, good ventilation should be provided. At a few seed-processing plants, cones of true firs and spruces are stored on ventilated mesh screens for artificial ripening.

There has been an attempt to rank the different species according to relative ability to withstand prolonged cone storage. The ranking is primarily based on intuition and experience, but such information is valuable in scheduling cone processing. Several studies conducted with western species have confirmed the success of current cone-storage practices and have shown that, if cones are handled properly and storage conditions are optimum, seed could be safely stored in intact cones for up to 4 to 6 months. For cone storage beyond 4 months, it may be advantageous to install frost protection because subfreezing temperatures could significantly reduce seed germinability.

Seed Processing

After cones are harvested and stored, seeds are extracted and prepared for either immediate sowing or storage. This series of operations, called seed processing, includes kiln drying, cone tumbling, scalping, dewinging, and cleaning and sorting.

Kiln Drying

Given good drying conditions, cones of most conifers open readily. Under natural storage conditions, however, cones may not be thoroughly and uniformly dried, especially when weather is humid and cool. Cones should therefore be kiln dried to facilitate extraction.

Kilns are of two types: rotating and progressive. In rotating kilns, a batch of cones is loaded into and dried within a drum where temperature and humidity are usually controlled. Such kilns, although not suitable for drying a large quantity of cones, can provide specific drying temperatures and relative humidities for small-lot processing. In progressive kilns, loaded trays are moved at certain time intervals to expose cones to increasingly warmer air as they dry. This type of kiln is more suitable for large-batch processing.

Kilns are generally operated at temperatures between 32 and 60°C. Although studies have shown that the biologically lethal temperature of most tree seed is around 66°C, the operational maximum temperature should not exceed 43°C. Because cones often have a high moisture content after storage, however, drying should be started at low temperatures that are progressively elevated. Drying cones with high moisture content immediately at high temperatures should be avoided because it could lead to case hardening and result in partial cone opening and incomplete seed extraction. However, the problem of case hardening can, to some extent, be overcome by moistening scales or soaking cones in water.

Air humidity is as important a factor as temperature. Low humidity is the key to more complete drying. For example, cones can be successfully dried at the relatively low temperature of 32°C if relative humidity is below 30%. Cones of most major conifers in the Northwest readily open upon drying. However, lodgepole pine cones from certain geographic areas are serotinous and require a short soak in hot water before kiln drying.

Additional soaking cycles with water have been reported to increase seed extraction by 20 to 84%.

Cone Tumbling

In rotating kilns, cones are dried and tumbled simultaneously, and seeds fall out as cones open. Generally, loose seeds drop through perforations in the drum. Cones dried in progressive kilns are subjected to shaking action by tumblers to extract seeds from cones. A tumbler is a rectangular or round wiremesh container mounted horizontally on its long axis, which turns at a slow speed. Small quantities of cones may be tumbled in batches. In large- scale continuous operation, the tumbler axis is inclined so that rate of cone movement through the tumbler can be regulated.

Scalping

Seeds coming from the tumbler must be separated from a mixture of cone fragments, hardened pitch, foliage, dust, and other debris. This step, called scalping, is achieved by vibration, air movement, or screens, alone or in combination. The most commonly used equipment has several layers of vibrating screens of different-sized mesh. Coarse materials such as scales and twigs are retained on the uppermost screen and slide down to be collected in one bin, while fine particles are screened to be deposited in another bin: the seed is usually collected through an intermediate screen.

Dewinging

Once debris has been eliminated, wings must be removed from many conifer seeds. Although wings are often loosened during tumbling and scalping, dry or wet dewinging may also be required. Dry dewinging, a technique which employs a rubbing action to remove wings from dry seed, is generally used for Douglas-fir, pines, and true firs. Small lots can be dewinged in a cloth bag; lots of up to 5 kg are better handled in a Dybvig macerator ; and large lots are best dewinged with a brush-type dewinger, although auger-type dewingers have also been used successfully.

Because dry dewinging is the processing step that is most likely to cause seed damage, extra caution should be exercised to use proper equipment and to minimize unnecessary friction. In one study, for example, three cycles of brushdewinging seeds of subalpine fir [*Abies lasiocarpa* (Hook.) *Nutt.*] destroyed 50% of the originally viable seeds. Because dry dewinging can mechanically damage seed, many seed workers prefer wet dewinging, especially for pines and spruces. The principle of wet dewinging is that wings are more hygroscopic than seed and, upon wetting, are released cleanly.

The Kason Vibrator and a rotating cement mixer with a soft brush have been successfully used for wet dewinging. However, because seed absorbs moisture during wet dewinging, it must be redried sufficiently before storage. Germination tests verified that a 20- to 30-minute water soak, followed by wet dewinging and air drying for 16 hours at 26 to 30°C to 4 to 8% moisture content, did not adversely affect seed quality.

Cleaning and Sorting

Empty seed, partially filled seed, and other foreign particles are removed from good seed in the final cleaning. Scalpers and fanning mills are often used for species that have few scales, such as Douglas-fir and pines, but vibratory gravity tables are best for true firs. Pneumatic seed cleaners have also been successfully used for various conifer species. All this equipment, in combination, further improves sorting efficiency. Flotation sorting with water, alcohols, and other organic liquids has been used to clean red spruce (*Picea rubens Sarg.*), true firs, and several pines, although this method has only been tested experimentally with western species.

A noteworthy development in seed sorting is the IDS (incubation-drying-separation) method, developed by Simak, which can separate nonviable, as well as empty and partially filled, seed from viable seed. Fully imbibed seed is first incubated for a short time, then gradually dried, and finally separated by various specific- gravity methods. Because empty and nonviable seeds lose water more quickly during the drying phase, differences between nonviable and viable seeds are magnified, making subsequent separation by standard gravity methods more effective. Scots pine (Pinus sylvestris L.) seed of low germinability was successfully upgraded by this method experimentally.

In US, most processing work is done at seed plants in the Northwest. All eight seed-processing plants responding to the OSU Nursery Survey process their own seed as well as seed harvested by other organizations. However, four of the 16 nurseries responding do at least part of the processing at their own facilities. The remaining nurseries have private or state plants process their seeds.

Seven of the eight seed-processing plants and seven of the 16 nurseries set their own standards of purity for commercial seed transactions or nursery sowing (Table 2). The seed plants and nurseries replying to the Survey had a generally higher standard of purity than the Western Forest Tree Seed Council recommendations for four of the six major coniferous species; the lower accepted purity standards of the true firs (Table 2) may indicate possible difficulties in removing nonseed components without adversely affecting seed germination.

Table 2. Minimum purity standards recommended by the Western Forest Tree Seed Council

Species	Tree Seed Council	Seed plants andnurseries
	%	%
Douglas-fir	95	95-99
Ponderosa pine	95	95-99
Lodgepole pine	90	99
Noble fir	95	90-98
White fir	95	90-98
Western hemlock	90	95

Seeds of true firs are known to be especially sensitive to handling and mechanical damage.

Seed Testing

Seed testing evaluates seedlot quality and is essential for both seedling production and commercial seed transactions. Most tree-seed tests are conducted with methods based on rules of the Association of Official Seed Analysts (AOSA) or the International Seed Testing Association (ISTA).

Sampling

The first step in seed testing is to draw a sample that represents the entire seedlot. A seedlot is defined as a unit of seed of reasonably uniform quality from a particular location or elevation. Seedlot size varies with testing rules and among laboratories. ISTA, for example, has determined that a seedlot should be less than 5,000 kg for seeds the size of beech (Fagus spp.) seed or larger, or 1,000 kg for seeds smaller than beech. The Western

Forest Tree Seed Council recommends that lots in excess of 227 kg be divided into equal smaller lots for sampling.

Loose seeds in containers should be sampled with seedsampling probes long enough to reach all areas in the containers. The sample should be composed of equal portions taken from evenly distributed volumes of the lots to be sampled, each sample proportional to the size of the container. Samples should be subdivided in the testing laboratory with a mechanical divider until a subsample of the desired weight is obtained.

Physical Characteristics

Purity

Purity tests measure the percent by weight of four major components: (1) pure seeds of the test species, (2) seeds of other crop species, (3) weed seeds, and (4) inert matter (leaves, cone scales, etc.). The purity test is usually the first test per—formed for a given lot and is especially important for commer—cial transactions, which are based on weight.

Moisture content

Seed moisture content is most often determined with the air-oven method. Seed samples are heated in ovens; the weight loss that occurs during drying is considered to be seed moisture. ISTA rules prescribe oven drying at 105°C for 16 hours for all tree seeds except those of the genera Abies, Cedrus, Fagus, Picea, Pinus, and Tsuga. Seeds of those genera contain a significant amount of volatile oils and resins which may be lost at the above temperature. Therefore, their moisture content must be determined by toluene distillation. Electronic moisture meters, though not as accurate as the above methods, are frequently used by various seed workers; they give rapid measurements desirable, for example, when checking moisture in a large number of seedlots being dried before storage.

Seed moisture content can be expressed as a percentage of water loss of either total fresh weight or corresponding ovendry weight. Seed moisture content has been expressed on a dry-weight basis in some research, but international usage is exclusively on the fresh-weight basis. To avoid misunderstandings, the base should always be clearly specified.

Weight

Seed weight, required for calculating sowing rates in nursery sowing and direct seeding, is a function of seed size, moisture content, and proportion of full seed in a given lot. The commonly used unit is the weight of 1,000 pure seeds (1,000 seed weight). ISTA specifies weighing eight random samples of 100 seeds each from the pure-seed component; however, some laboratories use two or more samples of 500 seeds each. When means of replicates vary more than 10%, additional samples should be weighed. All weights should be accurate to three significant digits.

Biological Characteristics

Tests to estimate seed viability

Germination potential, perhaps the most important quality measurement in seed testing, is used to determine sowing rates as well as whether seed must be sown immediately or can be stored. Seeds of different species have different requirements for optimum germination. This potential can be (1) evaluated directly by germinating seeds under predetermined conditions or (2) estimated indirectly with biochemical staining, embryo excision, cutting tests, x-ray radiography, or hydrogen peroxide tests.

The most reliable method is germination in a controlled environment. At least 400 seeds, usually divided into four replicates of 100 seeds each, from the pure-seed component of the purity test are normally prechilled for up to 28 days and germinated on suitable substrates. Substrates should (1) be nontoxic, (2) be free of molds or other microorganisms, and (3) provide adequate aeration and moisture ; those recommended by AOSA are blotter papers, paper towels, washed sand, vermiculite, perlite, and peat moss.

Most (over 70%) of the coniferous species listed in the AOSA rules are germinated under alternating temperatures (30°C for 8 hours in the light, 20°C for 16 hours in the dark). An intensity of 750 to 1,500 (± 250) lux [75 to 150 (± 25) foot—candles] is recommended. Seed is counted as germinated when all essential structures appear normal. Retests are necessary when an extremely high proportion of full, ungerminated seed is left at the end of the test, or when variation among test replicates exceeds the accepted tolerances.

Although controlled-environment germination tests are reliable, they are often time consuming, especially for dormant species requiring prechilling. Several rapid methods of estimat—ing viability have been proposed, two of which-tetrazolium staining and embryo excision-are now recognized as official testing procedures.

The tetrazolium test is the most commonly practiced biochemical staining method. Seeds are immersed in 2,3,5-triphenyl tetrazolium chloride. Living cells stain red as tetrazolium is reduced by dehydrogenase enzymes to form a stable red triphenyl formazan, which is insoluble in water. The method is fast but lacks uniformity in staining ; therefore, results can be difficult to interpret. Other biochemical staining methods applied to seed testing with varying degrees of success include those using salts of selenium and tellurium and Indigo Carmine.

The excised embryo test is recommended for several species of pines including Coulter pine (*Pinus coulteri D.* Don), Jeffrey pine, and sugar pine (*Pinus lambertiana Dougl.*). Excised embryos are cultured on moist filter or blotter paper in covered dishes under light for 10 to 14 days at 18 to 20°C. Viable embryos remain firm and white and turn green, indicating growth, whereas dead ones turn dark or are covered with mold. This method is fast but requires skilled analysts.

Other quick methods include the cut, x-ray, and hydrogen peroxide tests. In the cut test, seed is bisected and then rated visually; this is the simplest but most unreliable method because distinguishing seeds damaged during handling and storage is very difficult. The x-ray test is fast, especially when Polaroid film is used, and development of contrast techniques has greatly expanded x-ray test capabilities. Disadvantages are difficulty in interpretation and relatively high equipment costs. The hydrogen peroxide (H_2O_2) test allows assessment of root growth in 1% H_2O_2. It is simpler to perform than t he excised embryo test and is more objective and easier to interpret than x-ray. However, as with other quick tests, it tends to overestimate viability, compared with germination tests.

Seed vigor

Nursery bed germination is usually slower and less complete than laboratory germination. Therefore, various laboratories have attempted to define and determine seed vigor to improve prediction of nursery germination. Three major groups of expressions have been proposed: (1) mathematical values based on standard laboratory test results, (2) germination under stressful conditions, and (3) biochemical testing. Mathematical expressions have been most widely tested.

They include the number of days required to attain a certain proportion of total germination, germination value, modified germination value, and the Weibull function. Germination under stressful conditions has been developed mainly for seeds of agricultural species; most widely used are the cold test for corn and the accelerated aging test for soybean. However, application of these tests or development of new procedures for tree seed has been rather limited. Biochemical tests have been tried to a limited extent; the few reported include tetrazolium staining and the GADA (glutamic acid decarboxylase activity) test.

Seed Storage

Irregular and often infrequent seed production by many of the major tree sp ecies necessitates seed storage-sometimes for several years-to maintain supplies through years of poor seed production. Because of this, considerable research has been carried out on seed storage. Storage is one area of forest -tree seed technology for which sufficient information is available for most species of interest.

Successful seed storage requires knowledge of the seed characteristics of different trees as well as of the factors influencing storage capacity, such as seed quality before storage, seed moisture content, and storage temperature and method.

Seed Longevity

The life span of seeds varies with species. Seeds are classified into three biological categories according to their life span under natural conditions: (1) microbiotic seeds (life span not exceeding 3 years), (2) mesobiotic seeds (life span from 3 to 15 years), and (3) macrobiotic seeds (life span from 15 to more than 100 years). Seeds of most conifers and hardwoods are microbiotic. Under regulated storage conditions, however, longevity of many tree seeds can be extended more than tenfold. For example, the viability of naturally dispersed seed of spruce and many pines extends only into the first growing season and, occasionally, into the second growing season.

Under subfreezing storage, seed viability of these same species can easily be maintained at high levels for 10 years or longer. Storage over 10 years is not usually required for seedling-production purposes but may become vital to future tree-breeding programs. Under optimum storage conditions, seed viability of certain trees might be maintained indefinitely, but the maximum potential for maintaining original seed viability has not yet been determined for most species.

Seed Quality

Seed quality has a significant impact on storage capability. Factors affecting quality are seed maturity, cone handling, and seed extraction and processing. Immature seeds are not only poor in germinability and liable to be further damaged by seed processing but also are difficult to store successfully. Overheating during extraction and damage caused by dewinging also have been found to adversely affect seed quality. Injured seeds are not suitable even for short-term storage because they have a high rate of respiration, undergo spontaneous heating, and deteriorate rather quickly.

Seed Moisture Content

Of all the factors influencing seed storage, moisture content may be the single most important one in maintaining germinability. Various researchers have demonstrated the detrimental effect of high seed moisture on tree-seed viability; increased rates of respiration and changes in carbohydrates and fats presumably cause seeds to use their food reserves. Excessively low seed-moisture content also may reduce storage capability. Some species, including Douglas-fir, can tolerate drying to 0% moisture content ; however, overdrying can destroy the monomolecular layers that protect against oxidation. Recommended seed-moisture content for storing Douglas-fir, ponderosa pine, lodgepole pine, and western hemlock is 6 to 9% (wet-weight basis); that for true firs is 9 to 12%. However, Danielson and Grabe showed that optimum moisture content for noble fir is also 6 to 9%.

Storage Temperature

The effect of storage temperature on the retention of treeseed viability has been thoroughly investigated. The general relationship between storage temperature and moisture content was described by Barton as follows: at a given moisture content, the higher the storage temperature, the faster the deterioration of seed viability: the lower the storage temperature, the greater the tolerance to high moisture content and the better the retention of viability. Some studies have shown that temperature slightly above or below freezing may be sufficient to prevent deterioration for short -term storage, but the retention of viability has generally been better when seeds were stored at -18°C, particularly for longer periods.

Storage Method

Tree seeds can be stored either wet or dry. Large seeds of hardwood species require moist conditions and are usually kept wet for short-term storage, whereas small seeds of conifer species, including all major conifers in the Northwest, are stored dry. Seed moisture content is controlled by storing properly dried seed in tightly closed containers or by regulating humidity in the storage area (as for many agricultural seeds); in the Northwest, dried seed is generally placed in closed containers, although some facilities do use humidityregulated storage rooms. These facilities, although costly, are effective in minimizing reabsorption of moisture by dried seed, especially in areas with humid climates.

The most frequently used storage containers are plastic bottles with screw tops, polyethylene bags, and fiberboard drums. More than 80% of the organizations surveyed store their seeds in rigid drums lined with polyethylene bags. This method is common in many areas because it is relatively inexpensive and effectively prevents uptake or loss of moisture by seed from the atmosphere. Of 16 organizations, nine use plastic bags as containers; five of those use I- to 6-mil plastic, and the remaining four use 7- to 8-mil plastic. Thin bags are subject to ripping but easier to handle when cold. However, plastic containers are not completely impermeable to moisture. Use of thicker materials may be desirable for seed requiring low seed-moisture content if external humidity is high and seeds are to be stored for a long period.

Retesting

Retesting is often recommended for seeds stored for a relatively long period (5 years or more). Even under ideal storage conditions, certain poor -quality seedlots rapidly lose their viability. Fourteen of 16 organizations conduct viability tests at 2- to 6-year intervals. Seed moisture content is retested by only three of the 16, however, probably indicating that moisture content does not fluctuate significantly under current storage conditions.

Even if seed moisture changes only minimally in storage, additional moisture could be introduced when seeds are withdrawn. Therefore, it is particularly important that sealed containers removed from cold storage be permitted to reach ambient temperatures before being opened to avoid condensation of water within the container. Sahlen and Bergsten found that temperature was completely equalized in the center of a 28-liter container, with walls 2.5 mm thick, 36 hours after the container was moved from a -16°C storage room to 22°C ambient temperature. To minimize repeated opening and resealing and to reduce storage space, the use of smallsized containers (10- to 25-kg capacity) has been recommended.

Seed Stratification

Tree seeds. unlike agricultural seeds, are in many cases characterized by deep dormancy. This is true for most Northwest conifers. Seeds of different species or different geographical origins often require different pretreatments and conditions for optimum germination. The most commonly used pretreat—ment to break dormancy is stratification-which usually is moist cold treatment for up to several months. Stratification is generally known to bring about changes in anatomy or physiology, including embryo growth, and in metabolism. Physiologically, breaking of dormancy has often been explained in terms of a shift in the inhibitor-stimulator balance. Presumably, although it may not directly affect the level of inhibitors, stratification could increase growth-stimulator levels, which would then counteract the effects of inhibitors in breaking dormancy.

Successful cold stratification requires: (1) proper moisture content, (2) low temperature, (3) adequate aeration, and (4) proper length of time. In practice, seed originally was stratified by placing it between moisture-holding media such as peat moss or sand in boxes, tanks, trays, and oth er suitable containers and maintaining it there under cold, moist conditions. Some nurseries use outdoor soil pits. More recently, stratification in polyethylene bags has become common at many nurseries and seed-processing plants. This method, called "naked stratification," requires no moisture-holding medium and less effort in preparing seed for subsequent sowing. Seed is soaked in water in containers lined with plastic or mesh bags, drained of excess water, and kept at low temperatures for a predetermined period of time; bags are often loosely fastened to allow aeration.

Water Soaking

The rate of water absorption varies among species. Most conifers require 1 to 2 days of soaking to achieve full imbibition. It has been suggested that warm water can speed up water absorption by seed, and that running water and aeration can improve oxygen availability; however, this has not yet been substantiated experimentally for Northwest conifers. During soaking, four organizations aerate water, whereas three use running water. These practices are probably beneficial, although the effectiveness should be determined for each species.

Temperature

After draining, seeds are stored in the fully imbibed state.

A few species, such as yew (Taxus spp.) and yellow-cedar [Chamaecyparis nootkatensis (D. Don) Spach], require storage at warm temperatures before cold storage: however, most coniferous species require low temperatures throughout. For loblolly pine (*Pinus taeda L.*), McLemore found optimum stratification temperature to be 10°C, but Robinson et al. reported that, for this same species, gradually increasing temperature over a 4-week period gave the best stratification results.

Temperatures above 5°C are not desirable because they increase the risk of overheating and subsequent deterioration, alt hough freezing temperatures also can damage seeds at high moisture content. Consequently, low temperatures of 2 to 5°C have been adopted as accepted operational practice in most cases. All except one organization use stratification temperatures between 1 and 5°C; the exception uses 0°C. Even in this case, however, embryos would not experience freezing due to their osmotic potential, which is lower than that of water. Premature germination during prolonged stratification can be minimized if seeds are held at 2°C, rather than 5°C, for both Douglas-fir and ponderosa pine.

Aeration

Aeration during stratification is necessary to supply oxygen for seed respiration and to allow carbon dioxide and heat to escape. Lack of aeration could therefore lead to deterioration of seed quality through buildup of toxic substances. The most commonly used technique is to leave a small air space at the neck of each bag in which seed is stratified and to massage the whole bag periodically. A few Northwest nurseries also use fine-meshed bags hung so that air may circulate.

Duration

Optimum stratification length varies among species and seedlots. In general, the longer the stratification period, the greater the rate of germination, especially under suboptimal germination temperatures. For this reason, seed destined for colder environments must be stratified long enough for quick and complete germination. However, prolonged stratification can cause seeds of some species to germinate prematurely; furthermore, vigor and total germination may be reduced if seeds are stratified for excessively long periods.

Premature germination sometimes occurs during prolonged stratification. Although premature germination is a serious concern in nurseries because the fragile seeds can be damaged in handling or during mechanical sowing, redrying and storing of stratified seed are possible. Danielson and Tanaka reported that ponderosa pine seed air-dried to 26% and Douglas-fir seed air-dried to 37% can be stored for 9 and 3 months respectively without losing the beneficial effect of stratification or having their viability adversely affected. Subsequently, Edwards tested the efficacy of surface-drying true fir seed after 1 month of stratification at saturation moisture, followed by 3 months of storage at 35% moisture content; this treatment not only prevented premature germination but also improved total germination and germination rate.

The exact mechanism behind the benefit of surface drying is not completely understood. It may be related to improved gaseous exchange brought about by removing the water film from the seed surface, which increases oxygen availability to the seed and facilitates the release of any accumulated toxic gases. Although the surface-drying technique provides the option of storing stratified seed for prolonged periods without losing stratification effects, the lower limit of seed moisture content should be determined for each species. Seeds can be stored safely below certain thresholds but seem to then require restratification after storage. This induction of secondary dormancy suggests that seed moisture and dormancy are closely related.

Other Treatments

Although stratification is an effective method to break dormancy, it is often time consuming. Past research has shown that hydrogen peroxide, gibberellic acid, ethylene, microwave irradiation, or osmotic agents can stimulate germination of conifer seed. However, these studies were usually conducted under optimum germination conditions and may not be effective under the suboptimal temperature conditions frequently encountered in the field in early spring when seed is sown at Northwest bareroot nurseries. Further work is required to develop a quick, effective method that would facilitate germination under a wide range of temperatures and that would either eliminate the need for stratification or shorten the stratification requirement.

7

Measures of Seedling Quality

The carbohydrates are a diverse group of compounds sharing common building blocks-the sugars. Within this group are hundreds of compounds and, in the case of the sugars. those compounds are rapidly interconverted. This complexity can be reduced, however, by grouping the carbohydrates according to their functions in the plant.

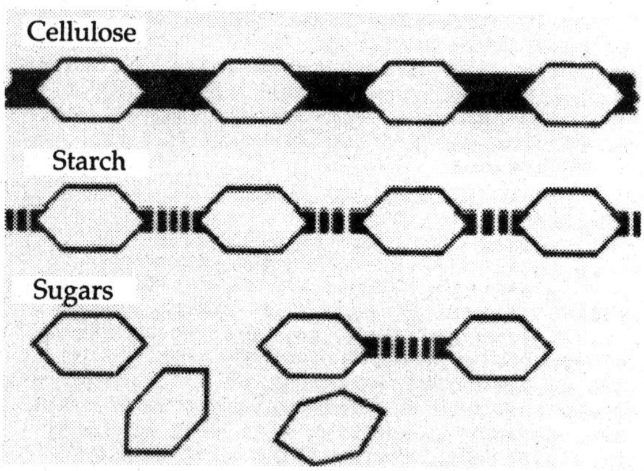

Figure 1. Stylised Structures of the major groups of Carbohydrates. Hexagons represent individual sugar molecules, and thickness of the bands connecting them indicates the strenfth of chemical bonds holding the sugars together

Structural Carbohydrates

In the seedling's carbon budget. the structural carbohydrates represent a non-refundable investment of carbon. Carbon spent by the plant to build structure is thereafter unavailable for other uses.

Cellulose is the predominant structural carbohydrate in trees. It is a linear chain (or polymer) of glucose molecules that cross-link with adjacent cellulose molecules. This cross-linkage gives the plant much of its strength. Cellulose. like other structural carbohydrates, is found almost exclusively outside the cell membrane in the cell walls.

The hemicelluloses are a second important group of structural carbohydrates occurring in the cell walls. They include complicated polymers consisting of a wide variety of sugars. There has been some speculation that hemicelluloses may at times serve as reserve materials in trees, but the evidence is inconclusive. Their exact function in the cell wall is not yet known. Finally. the pectins. which are chains of modified and unmodified galacturonic acid molecules, appear in the middle lamellae between cells. They serve as an intercellular glue and are apparently not used as reserve substances.

Reserve Carbohydrate

Starch is the most important reserve carbohydrate. It comprises the "savings account" from which a seedling can draw if respiratory or growth expenses exceed photosynthetic income.

Like cellulose. starch is composed of joined glucose molecules. Unlike cellulose, the starch molecule is easily disassembled by enzymes, a process that releases its constituent sugars. Starch occurs within special structures- amyloplasts and chloroplasts-inside the cells. In a rice grain or potato tuber. starch concentrations may reach 70% of the dry weight. In conifer needles and fine roots. starch concentrations may reach 30% of the dry weight. On the other hand. starch concentrations may drop to near zero in a root or leaf about to be shed.

Sugars

The sugars are the metabolically active carbohydrates and are involved in all biosynthetic and maintenance activities. They are the only form of carbohydrate transported from place the place within the plant. They can be regarded as the seedling's "pocket money." The sugars also serve as the major carbohydrate for storage when temperatures are low. Any starch present is likely to be converted to sugar during cold-hardening because the sugar acts as an antifreeze within the cells.

Carbon Inputs and Outputs

Photosynthesis

Photosynthesis is the process whereby CO_2 is drawn from the atmosphere and converted to simple sugars. The energy source for this process is sunlight. Photosynthesis represents the plant's income, its paycheck. All other chemical reactions within the plant depend upon the "burning" of these sugars to release energy. This burning of sugars, releasing CO_2, is what we mean by respiration. Respiration can be separated into that required for construction of tissue and that associated with maintenance of tissue. The relationships between each of these processes and the various kinds of carbohydrates are shown in Figure 2.

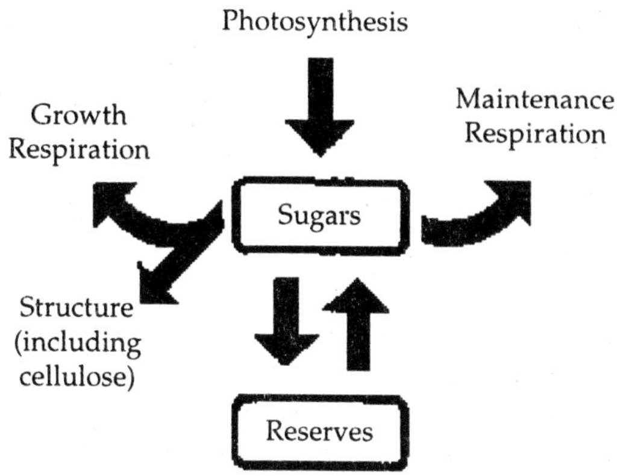

Figure 2. Relationship between the various types of Carbohydrates and the Processes of Growth and maintenance respiration

Respiration Growth

The seedling has two costs associated with growing new tissue. The greatest cost is that of the carbon actually being placed into the new tissue-for example, the sugars that are placed into cellulose polymers. The second cost is for the sugars burned to provide energy for synthesis of new molecules. In the case of cellulose, this is simply a matter of joining sugars. To make more complicated compounds, however, requires considerably more energy. The CO_2 released in these processes is attributed to "growth respiration."

The rate of growth respiration depends upon the growth rate and the biochemical composition of the tissue. Note that by this definition, growth is an irreversible commitment of resources. Because it is not irreversible, an increase in weight as a result of starch accumulation is not growth: the starch can be burned and the weight lost.

Respiration Maintenance

Respiration rates never fall to zero in living tissue; there is a cost associated with tissue maintenance. This cost is usually termed "maintenance respiration." It is associated with the burning of sugars to yield the energy needed for maintenance processes. These processes include replacement and repair of enzymes and cell membranes as well as transfer of materials from cell to cell. Because enzymes and membranes are replaced more fre-quently at higher temperatures, maintenance respiration increases exponentially with increasing temperature, approximately doubling with every 10°C increase.

Accumulation and use of Reserves

The seedling's only source of sugars, its only income, is through photosynthesis. The seedling does not necessarily use its sugars immediately, however. In fact, a large proportion of the sugar is often stored as reserves. The exact amount being stored depends in large measure upon the rates of photosynthesis and respiration. Therefore, reserve accumulation is favored by anything that (1) increases photosynthesis, (2) decreases growth, or (3) decreases rates of maintenance respiration. Alternatively, any of the opposite effects will reduce reserve accumulation and eventually begin to deplete reserves. Changes in light intensity, temperature. nutrition, or growth rates will alter these three major processes, resulting in predictable changes in reserves, as discussed in the following section.

Light

Within a limited range, photosynthesis is sharply increased by increasing irradiance. If such an increase does not cause an equivalent increase in growth, starch will accumulate. Alternatively. placing a seedling in darkness causes it to deplete its reserves, especially if growth continues. Only by using reserves accumulated during the day can seedlings continue to grow at night. High seedbed densities necessarily reduce the amount of light available to each seedling and may reduce starch and sugar reserves.

Plant Moisture

As plant moisture stress develops, growth is reduced long before there is any reduction in photosynthesis. Consequently, starch concentrations increase sharplv under mild moisture stress. Of course, once a drought becomes severe, photosynthesis is also halted. Accumulation of starch under mild water stress provides a means by which a plant may better survive a period in which photosynthesis has been halted.

Nutrition

In some species, mild nutrient deficiencies also reduce growth more than they reduce photosynthesis. New growth requires major nutrient inputs, whereas photosynthesis can continue with nutrients already in place. For these reasons. a decline in nutrient availability will result in starch accumulation. Improved nutrient availability, on the other hand. may increase growth rate and reduce accumulation of reserves.

Temperature Effects

Temperature affects photosynthesis as well as growth and maintenance respiration. Photosynthesis and growth peak at temperatures around 15°-25°C; however, rates of maintenance respiration continue to rise exponentially to about 45°C. Low temperatures can kill rapidly growing tissues and may also disrupt the photosynthetic apparatus in the foliage.

Seasonal Growth Patterns

The periods of growth of many tree species are controlled by daylength or chilling requirements rather than by resource limitations. Analysis of the processes discussed earlier nonetheless applies; starch and sugar accumulate before and after periods of growth for as long as photosynthesis is active. This accumulation explains the increases in starch concentrations commonly observed in the spring and late summer and the increase in sugar concentrations commonly observed in autumn.

Determining Carbohydrate Content Procedures

The carbohydrates comprise a complex group of compounds. To a certain extent, they can be analysed according to these functional groups as well.

Iodine Staining

Iodine staining of starch in intact tissue has been suggested as a measure of seedling quality ever since the subject of physiological grades was first introduced. Hellmers showed almost complete depletion of starch in shoots kept in cold storage for 4-1/2 months. Given its simplicity and low cost, this technique should be extremely useful as long as starch is the compound of interest.

Unfortunately, sugars do not stain with iodine: consequently, it is difficult to know whether a decrease in starch concentration means that starch is being respired or being converted to sugars. Nonetheless, iodine may well provide the most useful qualitative measure of carbohydrate status in the field. Iodine staining would be of little use, however, during the winter, when starch is converted to sugars. Because conversion to sugars occurs to a lesser extent in roots than in shoots, iodine staining might be useful even during the winter if applied to root tissues.

Extractions

Determination of the amounts of starch and sugar in a tissue requires that the starch and sugars be completely extracted. Because such extraction has proven difficult in the past, estimates of starch and sugar concentrations have varied widely. Sugars can be extracted with ethanol, methanol, or methanol:chloroform:water (MCW). Hot 80% ethanol has been most usual. The alcohol extract contains phenols, organic acids, and pigments as well as the sugars. It is necessary to remove these contaminants to determine the sugar concentrations colorimetrically.

The non polar pigments (including chlorophyll) and the phenols can be partially removed by a 100% acetone extraction prior to the ethanol extraction. The remaining non polar contaminants can then be removed by adsorbing them out of the ethanol extract with insoluble polyvinylpyrollidone. Organic acids can then be removed by precipitation with neutral lead acetate; a clear, nearly colorless extract is left.

Solvent fractionation (addition of chloroform to MCW extract) may also be used to rid the extract of non-polar contaminants. After the soluble sugars have been removed. starch can be extracted from the residue. Enzymes are often used in this extraction because of their specificity. Unfortunately, the starch-extraction enzymes are easily deactivated by any phenols remaining in the tissue or in solution and therefore may not yield complete extractions.

This problem is apparently much more severe in conifer tissues than in the crop plants for which most of these methods were developed. Haissig and Dickson may have circumvented this problem by extracting with MCW prior to starch extraction. but their method has not been rigorously checked for completeness of extraction.

Starch can also be extracted by acids, usually perchloric or sulfuric. Acid concentrations should be chosen carefully, however, as perchloric acid concentrations used in the past were recently demonstrated to hydrolyse cellulose. Prolonged extractions with 35% perchloric acid will provide complete extraction of all starch. This extraction also removes a small amount of structural carbohydrate, which should be corrected for. The correction can be made by precipitating the starch out of solution with iodine. by measuring the carbohydrate concentrations of the extract at the point at which iodine-staining disappears, or by allowing the tissue to exhaust its reserves by storing it in darkness at high temperature.

Non-specific Carbohydrate Determinations

Total hexose sugar concentrations of sugar extracts and starch extracts can be determined with a spectrophotometer by measuring color developed in the reaction with anthrone. The major hexose sugars appear to yield very similar color development per glucose equivalent.

Specific Carbohydrate Determinations

Finally, if one wants to identify specific sugars in the extracts, it is necessary to use chromatographic techniques. Alternatively. glucose concentrations can be determined from color development after adding glucose oxidase to the solutions. Given the ready interconversions among sugars and the utility of the anthrone reaction, however. there seems little point in trying to identify specific sugars in assessing seedling quality.

Predictive Abilities

Correlations between survival and carbohydrate status have been demonstrated for many species, but this relationship has not been found consistently. There are a number of problems in interpreting the literature. First, the diverse carbohydrate extractions and the analyses used in these studies vary in completeness of extraction and in what carbohydrates they extract. Second. differences in growth conditions, handling, storage, and planting conditions could easily induce or alleviate carbohydrate stress in the seedlings. Just as one would not expect a fertiliser response if the added nutrients were not limiting growth, so one should not expect a carbohydrate response if carbohydrate status were adequate. In the following section discuss, how carbohydrate status affects seedling growth and survival.

Potentials of Root qrowth

It has long been recognised that root growth is essential to the survival of the planted seedling, and a specialised literature has developed around the measurement of "root growth potential" (RGP) of seedlings. It was long thought that the major effect of improved carbohydrate status was an increase in RGP: however. Ritchie has recently presented strong evidence against this hypothesis. He measured a steady decline in "non-structural carbohydrates" (presumably starch and sugars) in seedlings in cold storage. although RGP rose sharply and then fell. He concluded that changes in RGP were not directly correlated with concomitant changes in concentrations of reserve carbohydrates. Many experiments involving girdling, shading, and defoliation suggest that root growth, at least in conifers, is dependent upon current photosynthate rather than reserves.

Respiratory Costs

If carbohydrate reserves are not related to RGP, then how else could they improve seedling growth and survival? Carbohydrate status presumably measures different aspects of seedling quality than does RGP. RGP, being closely related to bud dormancy, seems to be a measure of readiness to grow. Carbohydrate status, on the other hand, may be more a measure of capacity to maintain tissue function under stress, i.e., to withstand periods when respiratory losses exceed photosynthetic income. If so. then the effects of carbohydrate status on growth and survival might appear only when seedlings are forced to draw heavily upon their reserves.

It should be possible to calculate the point at which the carbohydrate balance "goes into the red" by using a process much like balancing a checkbook. Starting with a measured starch and sugar concentration, respiratory expenditures could be subtracted and photosynthetic deposits could be added as the seedling goes through its lifting, storage, and planting cycle. Moreover, it should be possible to work in the opposite direction, to calculate the total respiratory costs a seedling is likely to incur, to define that total as a threshold concentration and then attempt to grow seedlings with carbohydrate reserves in excess of the threshold. Unfortunately, few measurements of the magnitude of these respiratory costs are currently available. Nonetheless.

A seedling enters the winter months having stored a certain amount of carbohydrate. This amount varies with growing conditions. One can assume that the planted seedling must wait about 3 weeks to achieve a positive carbon balance. This delay occurs because the soil is cold, a state that limits water uptake. and because the photosynthetic apparatus has been disorganised by storage and by any frosts that may occur. During this time, the seedling must continue to meet its maintenance respiration costs, an activity that brings the reserve carbohydrate concentration down to 13%. In addition, the seedling must invest a large amount of photosynthate in preparing leaves for photosynthesis and buds for shoot elongation. This investment may require as much as 5 mg of carbohydrate per day for perhaps a week.

Long-term Effects

Carbon debts seem to carry over from one developmental phase to the next and from one growing season to the next. Thus, Engelmann spruce seedlings planted with poor carbohydrate status suffered high mortality the following winter because they were unable to develop sufficient cold-hardiness. In a Swedish study, Scots pine seedlings outplanted late in summer with reduced carbohydrate status still had significantly reduced growth the third growing season after planting.

It is also reasonable to expect that drought resistance will be reduced as a result of low carbohydrate status. Ritchie suggested that such reductions might occur because of an inability to adjust osmotically and retain positive cell turgor at low xylem water potentials.

8

Seed Moisture Content

The moisture content is the amount of water in the seed and is usually expressed as a percentage. It can he expressed on either a wet weight basis (where it is expressed as a percentage of the fresh weight of the seed) or on a dry weight basis (where it is expressed as a percentage of the dry weight of the seed).

The seed moisture content can either be accurately determined experimentally by scientific techniques, or it can be predicted approximately from the information available. The determination is destructive to the seeds used and since in many cases a prediction is sufficient, determination should only be carried out where essential.

A small change in seed moisture content has a large effect on the storage life of the seeds. Therefore it is important to know the moisture content in order to make a reasonably accurate prediction of the possible storage life of each accession.

When should moisture content be determined and when should it be predicted? Accurate determinations should be kept to a minimum because for many activities a prediction is sufficient. In most cases this means one accurate determination after the drying period, so that the moisture content of the stored seeds is known and can be used to make better predictions of storage life.

How should moisture content be determined and how should it be predicted? The methods for determination are modified from those of the International Seed Testing Association (ISTA) as recommended by the IBPGR Advisory Committee on Seed Storage. Other methods are available and can also be used provided that the results give an accurate determination of moisture content and are calculated on a wet weight basis, to enable comparisons to be made between materials and between genebanks.

The possibility of making reasonably accurate predictions of moisture content will improve as you gain experience with seeds of different species and use your knowledge of the origin of the seeds and the conditions in which they have been stored. The moisture content of a seed will equilibrate with the relative humidity of the air surrounding it. This moisture content is called the equilibrium moisture content and because it is constant for any species at known temperature and relative humidity, it can be used as an approximation of the actual seed moisture content.

Moisture contents will differ from seed to seed within the same accession, but should not be sufficiently different to make more than about 1% difference in moisture contents between seeds of the same accession and different accessions of the same species. This is sufficiently close for a prediction.

A list of a few common species and their equilibrium moisture contents is provided for guidance (Table 1). Further information is available from Cromarty, Ellis and Roberts. For species not on the list, it is possible to experimentally determine equilibrium moisture contents for these species at the temperatures and relative humidities at which you are working and prepare your own table to use for future predictions.

Determination of Seed Moisture Content

Determination of seed equilibrium moisture content through the following steps:

1. Do not waste seeds of germplasm for this, but use either excess seeds or those which are being discarded because they have lost viability.
2. Take two lots of seeds of the species of known weight and leave them in open containers in the environment of your genebank to equilibrate. This may be several days for small seeds and longer for large seeds. Seeds will have reached equilibrium with the moisture content of the air when there is no change in weight on different days.
3. Measure the relative humidity of the atmosphere where the seeds have been held for the past few days by using an aspirated hygrometer.
4. Measure the mean temperature.
5. Remove a sample of seeds and do an accurate determination of moisture content for each seed lot.
6. The mean of the two tests can be used as a guide because other seed lots of the same species should equilibrate at a similar moisture content under these conditions.
7. Repeat the determination the following day to make sure that the moisture content is stable and that the equilibrium has been reached,
8. The work can be repeated using a number of species which are of interest at a range of relative

humidities that are common in your country at different times of the year.

9. Make a table of the equilibrium moisture contents at different temperatures and relative humidities and use this to refer to in the future.

Table 1. Equilibrium moisture contents of some common crop seeds at 25°C[1]

Species	Percentage relative humidity					
	15	30	45	60	75	90
Barley (*Hordeum*)	6.0	8.4	10.0	12.1	14.4	19.5
Beet (*Beta*)		5.8	7.6	9.4	11.2	
Buckwheat (*Fagopyrum*)	6.7	9.1	10.8	12.7	15.0	19.1
Cabbage (*Brassica*)		5.4	6.4	7.6	9.6	
Carrot (*Daucus*)		6.8	7.9	9.2	11.6	
Cucumber (*Cucumis*)		5.6	7.1	8.4	10.1	
Egg plant (*Solanum*)		6.3	8.0	9.8	11.9	
Flax (*Linum*)	4.4	5.6	6.3	7.9	10.0	15.2
Groundnut (*Arachis*)	2.6	4.2	5.6		9.8	13.0
Lettuce (*Lactuca*)		5.1	5.9	7.1	9.6	
Lima bean (*Phaseolus*)		7.7	9.2	11.0	13.8	
Maize (*Zea*)	6.6	8.4	10.2	12.7	14.4	18.8
Mustard (*Brassica*)		4.6	6.3	7.8	9.4	
Oat (*Avena*)	5.7	8.0	9.6	11.8	13.8	18.5
Okra (*Abelmoschus*)		8.3	10.0	11.2	13.1	
Onion (*Allium*)		8.0	9.5	11.2	13.4	
Radish (*Raphanus*)		5.1	6.8	8.3	10.2	
Rice (*Oryza*)	5.6	7.9	9.8	11.8	14.0	17.6
Rye (*Secale*)	7.0	8.7	10.5	12.2	14.8	20.6
Sorghum (*Sorghum*)	6.4	8.6	10.5	12.0	15.2	18.8
Soyabean (*Glycine*)	4.3	6.5	7.4	9.3	13.1	18.8
Tomato (*Lycopersicon*)		6.3	7.8	9.2	11.1	
Turnip (*Brassica*)		5.1	6.3	7.4	9.0	
Watermelon (*Citrullus*)		5.1	6.3	7.4	9.0	
Wheat (*Triticum*)	6.5	8.5	10.4	12.1	14.6	19.8
Winter squash (*Cucurbita*)		5.6	7.4	9.0	10.8	

How to Predict Seed Moisture Content

Step 1. Measure environmental conditions

1. Measure the relative humidity of the atmosphere where the seeds have been held for the past few days by using an aspirated hygrometer.
2. Measure the mean temperature where the seeds have been held for the past few days.
3. If the seeds are from a different environment, they must be allowed to equilibrate with the current one before any predictions are made.

Step 2. Check the equilibrium moisture content in the table

1. Find the name of the crop in the left hand column of the tables of equilibrium moisture contents.
2. Read off the equilibrium seed moisture content at the relative humidity measured in Step 1 above.
3. Use this figure of approximate moisture content as your predicted value in making estimates of drying period or when an exact value of moisture content is not necessary.

Seed moisture content can be experimentally determined by various methods. The IBPGR Advisory Committee on Seed Storage have advised that, for use in genebanks, the ISTA methods are accurate and acceptable. However, considering that the cost of germplasm is high and moisture content determination is destructive, it has been recommended that the ISTA methods should be modified by reducing the weight of seeds required for each replicate whilst increasing the accuracy of the weighings.

The ISTA has only recommended standard methods for common crop seeds (Table 2). There are therefore no standard methods for many of the species stored in genebanks. The ISTA methods can be applied to seeds of these other species where appropriate as explained in the following pages.

Table 2. Species whose moisture content should be determined by ISTA methods

Modified low constant temperature oven method
Brassicas
Castor (*Ricinus*)*
Pepper (*Capsicum*)
Cotton (*Gossypium*)*
Eggplant (*Solanum*)
Falseflax (*Camelina*)
Flax (*Linum*)
Groundnut (*Arachis*)*
Onion (*Allium*)
Radish (*Raphanus*)
Sesame (*Sesamum*)
Soyabean (*Glycine*)*
All tree species
Modified high constant temperature oven method
Alfalfa (*Medicago*)
Asparagus (*Asparagus*)
Barley (*Hordeum*)*
Bean (*Phaseolus*)*
Beet (*Beta*)
Bentgrass (*Agrostis*)
Bermuda grass (*Cynodon*)
Black salsify (*Scorzonera*)
Bluegrass (*Poa*)
Brome (*Bromus*)
Buckwheat (*Fagopyrum*)*
Canarygrass (*Phalaris*)
Caraway (*Carum*)
Carrot (*Daucus*)
Chervil (*Anthriscus*)
Chickory (*Cichorium*)
Chickpea (*Cicer*)*
Clover (*Trifolium*)
Cocksfoot (*Dactylis*)
Cress (*Lepidium*)

Contd....

Contd....

Crested dogtail (*Cynosurus*)
Cucumber (*Cucumis*)
Cumin (*Curinum*)
Dallisgrass (*Paspalum*)
Fescue (*Festuca*)
Foxtail (*Alopecurus*)
Lettuce (*Lactuca*)
Lupin (*Lupinus*)*
Maize (*Zea*)*
Millet (*Panicum*)
Oat (*Avena*)*
Parsley (*Petroselinum*)
Pea (*Pisum*)*
Rhodes grass (*Chloris*)
Rice (Oryza)*
Rye (*Secale*)*
Ryegrass (*Lolium*)
Sainfoin (*Onobrychis*)
Serradella (*Ornithopus*)
Sorghum (*Sorghum*)*
Squash (*Cucurbita*)
Sweetclover (*Melilotus*)
Tall oatgrass (*Arrhenatherum*)
Timothy grass (*Phleum*)
Tomato (*Lycopersicon*)
Trefoil (*Lotus*)
Tufted hairgrass (*Deschampsia*)
Velvetgrass (*Holcus*)
Vetch (*Vicia*)*
Watermelon (*Citrullus*)*
Wheat (*Triticum*)*

* Indicates grinding is required for moisture content determination

How to Determine Seed Moisture Content

Step 1. Determine How may Seed are Available for the Test

1. The total weight of seeds used for a moisture content determination should be sufficient to make the test accurate and yet not be wasteful of seeds. The sample should be fully representative of the accession and a minimum of ten seeds should be used to prepare the samples for the test.

2. The ISTA rules for seed testing recommend that two replicates of 4 g of seeds are used for the determination of moisture content for official seed testing. However, this weight of seeds is not appropriate for use in genebanks because large numbers of small seeds are needed, unnecessarily depleting the accession sample size. Therefore, it is suggested that approximately 0.5 g of seeds per replicate should be used for the moisture content determination of seeds in genebanks.
3. The lower the weight of seed used, the more accuracy is required to achieve a true result. Small samples should be weighed with an analytical balance to four decimal places using light-weight dishes, so that the ratio of the weight of the seeds and the dish is not too disproportionate.

Great care is required especially with small samples when doing the determination to ensure that the results are valid. In order to reduce experimental error, light-weight dishes can be made from strong aluminium foil, but these are fragile to handle and covers are not easy to make.

Always work with care and finish one sample at a time. Do not leave the dishes open in the laboratory between weighings because the seeds will either lose or absorb water from the air and small changes in weights can result in large differences in the calculations when the amount of seed used is small.

Step 2. Determine if Pre-drying if Pre-drying is Required

1. Use your experience to predict moisture content as explained in the previous Section.
2. If the seeds are very moist, pre-drying may be necessary. ISTA requires pre-drying for seeds of the species which require grinding listed in Table 2 of

more than 17% seed moisture content, for seeds of soyabean of more than 10% moisture content, seeds of rice of more than 13% moisture content and seeds of other species of more than 30% moisture content.

Step 3. Deterimine if Grinding is Necessary

1. Check the ISTA list of species to see if the seeds require grinding (Table 2).
2. If the seeds are not on the list but are larger than wheat seeds, grinding is required.

Step 4. Decide which is the Appropriate Method to Use

1. Two methods are recommended: one for oily seeds and one for all others.
2. If there is a recommended ISTA method for seeds of this species (Table), use the modified version of this method.
3. If the species is not included in Table 2, but has oily seeds use the modified ISTA Low Constant Temperature Oven Method.

Step 5. Proceed with the Determination of Moisture Content

The standard ISTA determination methods have been modified for use in genebanks in this manual by decreasing the weight of seeds required for each replicate and the use of balances which weigh to more decimal places.

Modified ISTA low constant temperature oven method

1. If pre-drying is not required, weigh out approximately the total weight of seeds to be used, as determined in Step 1.

2. If pre-drying is required, weigh out accurately two sub-samples of between 1 - 1.5 g of seeds. Pre-dry these samples in a warm dry place overnight. Weigh again, writing all the weights in the note book.
3. If grinding is necessary, grind the seeds of each sample, directly before weighing, into pieces not larger than 4 mm for legumes and tree seeds and 0.5 - 1 mm for cereals.
4. Keep the ground seeds in a covered dish and go directly to the weighings.
5. Weigh one clean numbered dish and cover accurately to 4 decimal places using an analytical balance. Write the weight in the notebook.
6. Add approximately 0.5 g of the prepared seeds distributed evenly over the base of the dish, replace the cover and accurately weigh the dish and cover and write this weight in the notebook.
7. Place the dish in a safe place and continue to do the second replicate and additional samples in the same way.
8. When all samples have been weighed into numbered dishes, place each dish on top of its numbered lid in the oven at 101 - 105 °C.
9. Wait for the oven to reach this temperature and heat the samples for 15 - 17 hours.
10. Remove the dishes from the oven, replace their covers and place in a desiccator to cool for 30 to 45 minutes at room temperature.
11. Remove the dishes one by one from the desiccator and immediately weigh each dish and cover and note the weight in the notebook. Do not leave the desiccator open during the weighings.

Modified ISTA high constant temperature oven method

1. If pre-drying is not required, weigh out approximately the total weight of seeds to be used, as determined in Step 1.
2. If pre-drying is required, weigh out accurately two sub-samples of between 1 - 1.5 g of seeds. Pre-dry these samples in a warm dry place overnight. Weigh again, writing all the weights in the note book.
3. If grinding is necessary, grind the seeds of each sample, directly before weighing, into pieces of not larger than 4 mm for legumes and tree seeds and 0.5 - 1 mm for cereals.
4. Keep the ground seeds in a covered dish and go directly to the weighings.
5. Weigh one clean numbered dish and cover accurately to 4 decimal places using an analytical balance. Write the weight in the notebook.
6. Add approximately 0.5 g of the prepared seeds distributed evenly over the base of the dish, replace the cover and accurately weigh the dish and cover and write this weight in the notebook.
7. Place the dish in a safe place and continue to do the second replicate and additional samples in the same way.
8. When all samples have been weighed into numbered dishes, place each dish on top of its numbered lid in the oven at 130 - 133 °C.
9. Wait for the oven to reach this temperature and heat the samples for 2 hours for cereals, 4 hours for maize and 1 hour for all other species on the ISTA list.

10. Remove the dishes from the oven, replace their covers and place in a desiccator to cool for 30 to 45 minutes at room temperature.
11. Remove the dishes one by one from the desiccator and immediately weigh each dish and cover and write the weight in the notebook. Do not leave the desiccator open during the weighings.

Step 6. Use the Results of the Weighing to Calculate Moisture Content

1. Moisture content is calculated on a wet weight basis and expressed to one decimal place.
2. Replicates should not differ by more than 0.2%.
3. For samples tested without pre-drying use the following formula:

$$\% \text{ Moisture Content} = \frac{\text{Weight of fresh seeds} - \text{Weight of dry seeds}}{\text{Weight of fresh seeds}} \times 100\%$$

4. For samples which have been pre-dried, use the following formula:

$$\text{Final \% moisture content} = S1 + S2 - \frac{S1 \times S2}{100} \%$$

where S1 = Percentage moisture content from first stage of drying and S2 = Percentage moisture content from second stage of drying

Step 7. Enter the Data into the Data Files

1. If the standard method was used, enter the percentage moisture content of each accession into the data file for that accession.

2. If the standard method was not used, also enter a note of the method used to determine it.

Example 1. One stage drying

Material	Dish number	Wt. of dish (g)	Wt. of dish and fresh seed (g)	Wt. of dish and dry seed (g)
C124 rep. 1	35	10.3245	14.8668	14.6356
C124 rep. 2	29	10.1442	14.9948	14.7485

Calculation:

Rep. 1:

$$\% \text{ moisture content} = \frac{14.8668 - 14.6356}{14.8668 - 10.3245} \times 100 = 5.1\%$$

Rep. 2:

$$\% \text{ moisture content} = \frac{14.9948 - 14.7485}{14.9948 - 10.1442} \times 100 = 5.1\%$$

Example 2. Two Stage Drying (Pre-drying)

Material	Dish number	Wt. of dish (g)	Wt. of dish and fresh seed (g)	Wt. of dish and dry seed (g)
First stage of drying (pre-drying)				
Z099 rep. 1	07	20.4112	35.6615	33.2419
Z099 rep. 2	08	21.3121	36.7446	34.2689
Second stage of drying				
Z099 rep. 1	61	10.2323	14.8839	14.6532
Z099 rep. 2	62	10.4132	14.9647	14.7475

Calculation:

Rep. 1

% moisture content from pre - drying (S1)

$$= \frac{35.6615 - 33.2419}{35.6615 - 20.4112} \times 100 = 15.9\%$$

$$\% \text{ moisture content from second drying (S2)} =$$

$$\frac{14.8839 - 14.6532}{14.9647 - 10.4132} \times 100 = 5.0\%$$

$$\text{Final \% moisture content} = 15.9 + 5.0 - \frac{15.9 \times 5}{100} = 20.0\%$$

Rep. 2

$$\% \text{ moisture content from pre-drying (S1)} =$$

$$= \frac{36.7446 - 34.2689}{36.7446 - 21.3121} \times 100 = 16.0\%$$

$$\% \text{ moisture content from second drying (S2)} =$$

$$\frac{14.9647 - 14.7475}{14.9647 - 10.4132} \times 100 = 4.8\%$$

9

Characteristics of Mineral Nutrition in Seedling

A quality seedling represents an integration of many morphological and physiological factors in much the same way that human health represents a multitude of interacting facets of human physiology. Mineral nutrition is just one of a number of physiological characteristics that contribute to a healthy seedling, but the relative importance of nutrition to seedling survival and growth after outplanting is hard to quantify. To those involved in reforestation, there is an intuitive relationship between mineral nutrients and seedling quality.

Tree seedlings, like all plants, are autotrophs that utilise solar energy to produce organic compounds from the raw materials of water, carbon dioxide, and mineral elements. While some mineral nutrients may become limiting for the forest seedling, the nursery manager is able to supply nutrients through fertilisation for optimal seedling growth. A tree nursery, therefore, should be able to produce high quality seedlings that are charged with optimum levels of mineral nutrients when delivered for outplanting.

Mineral nutrition can have both positive and negative effects on seedling quality, however. Large, well-balanced seedlings with healthy, green foliage should logically exhibit better outplanting performance compared to stunted, chlorotic seedlings. oWithholding nutrients such as nitrogen resulted in chlorotic seedlings that were half the size of fertilised ones. Seedling size has been shown to be positively correlated with foliar nitrogen which was in turn directly related to nursery fertilisation.

Rules of Mineral Nutrition of Tree Seedlings

More than half the elements in the periodic table have been found in plant tissue because most chemical ions in the soil solution are passively absorbed in the large volume of water that is absorbed during transpirational uptake. Only 16 elements have been proven to be required for plant growth. A mineral nutrient must meet two criteria if it is to be considered essential for plant growth.

First, it must be required for the plant to complete its life cycle, and second, it must be part of some plant constituent or metabolite. Of these 16 essential nutrients, carbon, hydrogen, and oxygen are obtained from water and carbon dioxide and together account for approximately 96% of the dry weight of plant tissue. The remaining 13 elements are of mineral origin, being absorbed as ions from the soil. Based on relative concentration, these elements have been divided into six macronutrients and seven micronutrients although the actual distinction is somewhat arbitrary (Table 1).

The 13 mineral nutrients in Table 1 are listed in order of relative concentration for general plant tissue, nitrogen being most abundant (1.5%) being followed by potassium (1.0%) and calcium (0.5%) and these values closely agree with the concentration levels reported for loblolly pine (Pinus taeda L.) seedlings.

Table 1. Concentrations of 16 essential elements in healthy plant tissue

Element		Proportion of O.D. Wt. %	Cum. %
1.	Carbon (C)	45	
2.	Oxygen (O)	45	
3.	Hydrogen (H)	6	96.0
	Macronutrients		
4.	Nitrogen (N)	1.5	
5.	Potassium (K)	1.0	
6.	Calcium (Ca)	0.5	
7.	Magnesium (Mg)	0.2	
8.	Phosphorus (P)	0.2	
9.	Sulfur (S)	0.1	3.5
	Micronutrients	ppm	
10.	Iron (Fe)	100	
11	Chlorine (Cl)	100	
12.	Manganese (Mn)	50	
13.	Zinc (Zn)	20	
14.	Boron (B)	20	
15.	Xopper (Cu)	6	
16.	Molybdenum (Mo)	0.1	0.5
			100.0

The functions of mineral elements vary from the structural components cf plant cells to the physiological actions of molecules such as enzymes. All the macronutrients, with the exception of potassium, are incorporated into cellular constituents (e.g., magnesium in the chlorophyll molecule) but may also serve physiological functions as coenzymes or enzyme activators. Micronutrients primarily serve in a variety of metabolic functions in cells but do not constitute a significant part of any structural component (Table 2).

Mineral Nutrient Uptake Patterns

The relationship between mineral nutrient uptake and

plant gro'vth follows a characteristic pattern. When a nutrient is present in relatively low concentrations in plant tissue, it is considered deficient and limiting to plant growth. At the lower ranges of this deficiency, the plant often exhibits certain observable characteristics and these deficiency symptoms can be helpful in diagnosis of the deficiency.

Table 2. Structural and Physiological functions of mineral nutrients in Plant Tissue

Mineral Nutrient	*Functions*
Macronutrients	
1. Nitrogen (N)	A constituent of amino acids and proteins but also found in a variety of other compounds such as chlorophyll, enzymes and cell membrances.
2. Phosphorus (P)	Gound in nucleoproteins, phospholipids and high-energy phosphate bonds which serve as the major energy transfer mechanism in cells.
3. Potassium (K)	Specific functions are unclear but serves osmotic functions in cells and may also be involved in enzyme activity. Not a constituent of any known organic compound.
4. Calcium (Ca)	Occurs as a major constituent of cell walls and is also involved in N metabolism. Also an activator of enzymes and is required for cell division.
5. Magnesium (Mg)	A structural component of the chlorophyll molecule and is also involved in enzyme function.
6. Sulfur (S)	A component of several amino acids, proteins and coenzymes.
Micronutrients	
7. iron (Fe)	Found in chloroplasts and serveral respiratory enzymes.
8. Manganese (Mn)	Essential for the synthesis of chlorophyll and also serves as an enzyme activator.

Contd....

Contd....

9. Zinc (Zn)	Involved in the synthesis of the hormone IAA and a constituent of serveral enzymes.
10. Chlorine (Cl)	Performs non-specific osmotic functions and acts in conjunction with enzymes.
11. Boron (B)	Plays a regulatory role in carbohydrate metabolism.
12. Copper (Cu)	A Constituent of serveral enzymes.
13. Molybdenum (Mo)	Involved in several enzyme systems including nitrate reduction.

At slightly higher concentrations, however, the deficient nutrient is still low enough to limit plant growth but not low enough to produce deficiency symptoms; this condition is called hidden hunger because it is difficult to visually diagnose. One interesting point in the nutrient uptake curve is called the Steenbjerq effect and occurs only when a nutrient is extremely deficient. When the limiting element is first supplied to the plant, a rapid increase in growth occurs which actually results in a temporary decrease in tissue nutrient concentration.

When the supply of the nutrient is no longer limiting to growth, the plant growth rate increases rapidly until the critical point is reached. The critical point is the tissue nutrient concentration at which the growth rate declines significantly and is usually defined as 95% of the maximum growth or yield. The range of nutrient concentration at which maximum growth occurs has been defined as the optimum range. Plants may continue to take up mineral nutrients even though this additional uptake does not result in more growth (luxury consumption). When tissue nutrient concentrations reach extremely high levels, toxicity can occur with certain elements because plant growth begins to decrease with additional amounts of nutrient.

Interactions between Nutrients

Nutrients are not absorbed or utilised independently of each other and any change in the concentration of one element is usually accompanied by changes in the concentration of others. A dilution effect occurs when an increase in the concentration of one limiting nutrient causes increased plant growth that results in a decrease in the tissue levels of other nutrients. Antagonisms between nutrients (e.g., excess ammonium nitrogen causing reduced uptake of potassium) can occur during absorption or within plant tissues. Nutrient interactions are particularly common among the micronutrients; a complex between levels of iron, manganese, zinc, and copper exists where the relative concentration of one element affects uptake and utilisation of all the others.

Affect Nursery Practices on Seedling

Tree seedling nurseries utilise many cultural procedures to enhance plant growth and many of these practices have an influence on seedling nutrition. Duryea and McClain discussed four of the most significant practices in bareroot nurseries: fertilisation, seedbed density, irrigation, and root culture. A fifth practice, lifting date, can be added.

Fertilisation

The most obvious way to influence seedling nutrient status is through fertilisation.

Application of organic or inorganic fertilisers to the nursery bed or container media will usually affect the nutrient status of the seedling even if the applied nutrient is not limiting to growth and luxury consumption occurs. Fertilisation with one nutrient may also affect the levels of other nutrients in plant tissue due to the dilution effect. In an experiment with nitrogen fertilisers on

Douglas-fir (Pseudotsuqa menziesii (Mirb.) Franco) seedlings, the N concentrations in the foliage increased but the P and K concentrations decreased.

Different species may react differently to fertilisation. Applications of N fertiliser to four different conifer seedlings increased foliar N concentration in coastal and interior Douglas-fir and Sitka spruce (Picea sitchensis (Bong.) Carr.) but not in lodgepole pine (Pinus contorta var. latifolia Engelm.) (Table 3). Different parts of the seedling may also react differently to fertilisation. Switzer and Nelson reported that N fertiliser applications increased the N concentration of foliage, stem, and roots whereas P and K fertilisers only affected foliar concentration of the respective nutrients.

Table 3. Effect of Nitrogen (N) Fertiliser Applications on Foliar N Concentrations of four seedling species

Fertiliser Treatement (Kg N/ha)	*Costal Douglasfir*	*Interior Douglasfir*	*Sitka Spruce*	*Lodgepole pine*
60	1.04a	1.30 a	1.07 a	1.71 a
140	1.46b	1.51b	1.53 b	1.67 a
235	1.64c	1.59b	1.75 c	1.69 a

Timing of fertiliser applications is also important. Late season applications of N increased foliar N concentration to luxury consumption levels in five conifer species and this fertiliser effect was also reflected in foliage color. Applications of K fertiliser did not produce as dramatic results but still resulted in high K concentrations in the foliage.

Seedbed density

Control of seedbed density is probably the second easiest way to affect seedling nutrient status because decreasing the number of trees per unit area should increase the amount of soil nutrients that are available to each seedling. Reducing seedbed densities from 1200 to 300 seedlings per m^2 resulted in up to a 3X increase in N, Ca, and Mg content; a 5X increase in P; and a 4X increase in K content. In a experiment that studied both seedbed density and fertilisation effects, both foliar N concentrations and color were significantly enhanced by decreasing density.

Table 4. Effect of seedbed spacing on foliar nitrogen concentrations of four seedling species

Seedbed Treatement (cm)	Foliar N Concentrations (% dry wt.) Costal Douglasfir	Interior Douglasfir	Sitka Spruce	Lodgepole pine
0.6	1.08a	1.35 a	1.11a	1.63 a
1.0	1.08a	1.29a	1.12a	1.71a
2.0	1.21ab	1.23a	1.20 ab	1.74 a
4.0	1.37 b	1.34 a	1.33 b	1.73 a
8.0	1.75 c	1.73 b	1.89 c	1.66 a
12.0	1.79 c	1.88 b	2.07 d	1.68 a

Tree species are affected differently by changes in seedbed density. van den Driessche found that coastal and interior Douglas-fir and Sitka spruce had greater foliar N concentration with increased space between seedlings whereas lodgepole pine showed no effect (Table 4).

Irrigation

Irrigation can affect seedling nutrient status both positively and negatively. Excessive irrigation or heavy rainfall can leach soluble nutrients such as nitrate-nitrogen and potassium from the root zone and reduce the amount available to the plant. Eastern white pine (Pinus strobus L.) seedlings showed reduced foliar concentrations of N, P, and K with greater frequency of irrigation but Ca and Mg levels were unaffected. On the other hand, moisture stress caused by reduced irrigation can limit seedling growth and change foliar nutrient levels. Droughtstressed loblolly pine seedlings had significantly higher N levels than seedlings receiving ample irrigation.

Decreased irrigation also resulted in higher foliar N levels in white spruce (Picea glauca (Moench) Voss).

Root wrenching

Wrenching bareroot seedlings by undercutting the root system with a horizontally-drawn, angled blade temporarily stresses seedlings by severing roots and disturbing the soil. Wrenching could have either a positive or negative effect on seedling nutrient status. Water and nutrient uptake may be initially reduced until the root system becomes reestablished however nutrient and water uptake should eventually increase due to e denser, more fibrous root system after a single wrenching.

Wrenching Douglas-fir seedlings at intervals during the growing season reduced foliar nutrient levels with P and K more affected than N. Menzies reported that wrenched Douglas-fir seedlings had significantly lower foliar levels of N, P, K, and Ca when lifted in December and that the levels of these nutrients declined significantly when March lifted. Repeated wrenching of radiata pine (Pinus radiata D. Don) seedlings resulted in critically low N and P levels in the foliage. All authors agreed that supplemental fertilisation may be required to compensate for the possibility of reduced nutrient uptake due to repeated wrenching.

Lifting date

The time of lifting will also affect seedling nutrient status through foliar leaching losses, reduced uptake due to cold soils, and the dilution effect of additional growth without a corresponding increase in nutrient uptake. The concentration of N, P, and K ın slash pine (Pinus elliottii Engelm.) decreased when sampled during November, January, or March liftings whereas Ca and Mg levels increased slightly although levels in loblolly pine were unaffected.

Table 5. Relationship between lifting date and mineral nutrient lelves in 2+0 Douglas-FIR seedling Foliage

Mineral Nutrient Concentration	*October 16*	*Lifting Data December 18*	*March 3*
N (%)	1.93 A	1.80 B	1.54 C
P(%)	0.15 A	0.14 B	0.13 C
K (%)	1.00 A	0.76 B	0.60 C
Ca (%)	0.41A	0.41 A	0.38 B
Seedling dry weight	6.4 A	7.8 B	7.9 B

In a study of two fall lifting dates for Norway spruce foliar concentratio rof all macronutrients were lower for the second lifting date except Mg which remained relatively constant and S which was not measured. Menzies found that foliar levels of N, P, K, Ca, and Mg of 2+0 Douglas-fir seedlings decreased significantly between December and March liftings. van den Driessche monitored seedling dry weight and foliar nutrient levels over 3 lifting dates from October to March and found that mineral nutrient concentration decreased in all cases whereas dry weight increased (Table 5).

Methods of Characterising Seedling Nutrient Status

A practical method of measuring and rating seedling nutrient status is needed before it would be possible to correlate mineral nutrition with seedling quality. Based on the literature there are three possibilities: foliar symptoms, nursery fertilisation, and chemical analysis of plant tissue.

Foliar symptoms

The color of seedling foliage is affected by a number of biotic and abiotic factors. Foliar symptoms have been useful in diagnosing nutrient deficiencies, and specific symptoms may be correlated with individual nutrients. Many nutrient deficiency symptoms such as chlorosis, however, are not specific for an individual nutrient. Although chlorophyll concentration has been shown to be positively related to nitrogen status of seedlings and foliage color has been classified according to Munsell color charts, color was not considered to be a reliable indicator of seedling grade in a review of grading criteria. Because of this lack of specificity, foliar symptoms would not be useful for predicting outplanting performance of tree seedlings except in extreme cases.

Nursery fertilisation

The practice of fertilisation has a definite effect on seedling nutrient status. Although at one time it was possible to find nurseries that did not use fertilisers, it is now an accepted and widespread cultural practice in bareroot and container nurseries. Fertiliser application is indirectly related to seedling nutrient status because the mere application of fertiliser does not insure that the nutrients will be taken up by the seedling. Many research studies have related fertilisation to outplanting performance, however, and have inferred seedling nutrient status by quantifying nursery fertilisation.

Chemical analysis of plant tissue

The best way to characterise the mineral nutrient status of seedlings is to chemically analyse plant tissue and measure the levels of the various nutrients. Analytical chemistry techniques have been developed that allow accurate and precise determination of each of the 13 mineral nutrients in a small sample of seedling tissue. Seedling nutrient analysis can be used for two purposes: first, to diagnose nutrient deficiencies at the nursery and improve fertilisation practices and second, to predict growth response either at the nursery or after outplanting.

Measuring the Chemical Composition

Seedling nutrient analysis can be performed by chemical testing laboratories which are either private or are located at state landgrant universities. Most of these facilities work primarily with agricultural or horticultural crops and have little experience with tree seedlings. Although the actual chemical testing methods are basically the same for any crop, it is recommended that testing be done by a lab that has experience dealing with tree nursery stock. Before any actual sampling or testing is begun, it would be wise to discuss sampling, handling, storage, and shipping with the laboratory so that potential problems are avoided.

Selection of Sample Material

As with any scientific test, sample selection is extremely important because the characteristics of this limited sample must be representative of the population at large. The chance for serious error is much greater during the sampling phase than during the laboratory phase due to the large amount of inherent variability between seedlings and nursery location. A sound sampling procedure for seedling tissue analysis is provided by Solan.

Type of tissue

Either whole plants or parts of plants can be used for chemical nutrient analysis and the type of tissue depends on the objective of the test. To determine total nutrient uptake for fertiliser studies, whole plants should be tested; Leaf stated that no good evidence exists for analysing only one plant component for determining nutrient status of the entire plant. In New Zealand, however, the common practice is to analyse only the shoot of 1+0 radiata pine seedlings and in Great Britain, the Forestry Commission recommends needle samples for nutrient analyses. The common practice in agriculture is to sample leaves or petioles at a certain physiological age and there is good evidence that foliar samples are most appropriate in forestry. The foliage is the site of photosynthetic activity and is therefore a sensitive indicator of mineral nutrient status, and therefore van den Driessche recommends sampling whole seedlings or shoots for 1+0 seedlings but only foliage in older stock. This also agrees with the recommendation of Youngberg in the Forest Nursery Manual.

The best tissue for nutrient analysis is dependent on whether you want to measure nutrient concentration or content. Because nutrient concentration is a proportional measure, either foliage or whole plants may be used. For total nutrient content, however, the entire seedling should be analysed or dry weights and tissue samples of each seedling part taken so that the total content can be calculated. Both nutrient concentration and content have been used for seedling quality testing but, on a practical basis, concentration is most useful.

Time of collection

Seedling nutrient level varies during the year due to differences in nutrient uptake and utilisation and because of the dilution effect caused by new growth. Sampling date is therefore more of concern to those involved with testing for fertility analysis during the growing season because samples for seedling quality testing would normally be collected at the time of lifting or during storage. Seedling nutrient status has been shown to vary during the lifting season, however, and so therefore samples should be collected at a standard time each season. Nutrient levels should not change appreciably during refrigerated storage because seedling respiration losses would be inconsequential and so acceptable samples could be obtained any time during this period.

Number of samples

The number of samples to collect depends on the variation in the population (e.g. species, seed source, cultural treatment differences) and is usually limited by cost. A complete nutrient analysis of plant tissue generally costs from $10-50 per sample depending on the specific tests requested and so the total number of tests can be established by dividing the funds available by the testing fee.

Tissue samples are generally submitted as composites of individual seedlings from the population of interest. A composite sample should consist of a minimum of 20 to 50 seedlings although only 10g of dry or 60g of fresh plant tissue are required for the actual analysis. Composite samples, however, tend to "averageout" extreme values that may have biological significance and so there are advantages to analysing individual seedlings if they are large enough to meet the minimum tissue requirement. The best procedure is to discuss the objective of the test with the analytical lab and base your sample size on their recommendations.

Handling, Storing, and Shipping Samples

Seedling samples should be clean and free from soil contamination which can affect nutrient levels, especially of micronutrients like iron. Brushing or wiping the foliage clean with a wet cloth may help remove surface dirt but the best recommendation is to obtain clean samples in the first place. Excessive washing of seedling foliage should be avoided at the nursery because of the possibility of leaching out soluble nutrients like potassium.

The seedlings should be placed in plastic or paper bags and legibly labeled. The bays should be kept out of direct sunlight and placed into an ice chest to minimise metabolic activity which can affect nutrient status if the sampling period exceeds 4 hours or the weather is warm.

Once all samples have been collected, the needles can be removed from the stem of larger seedlings either manually or clipped with scissors. Although some sources recommend ovendrying at 65-80°C to stop metabolic activity, some N may be lost at temperatures as low as 70°C. The best procedure is to freeze the tissue samples immediately which achieves the desired result without danger of volatilisation losses.

To prepare the tissue for shipping, the frozen samples should be packed into Styrofoam ice chests or insulated cardboard cartons along with dry ice or packets of "blue ice". The shipping parcel should be sealed well and shipped by air freight or other reliable method of transportation so that they will arrive within 4-5 days. Again, the best recommendation is to check with your analytical laboratory to determine the best handling procedures.

Preparing Samples

Once the samples are received at the laboratory, they are processed through a series of steps to prepare them for chemical analysis including washing, drying, grinding, and screening.

Washing

Fresh plant samples are usually washed at the lab to reduce soil contamination but dried samples should never be washed. Washing the foliage in a dilute (0.1-0.3%) detergent solution followed by a distilled water rinse is satisfactory if done quickly to avoid leaching out nutrients. Samples can also be cleaned with an ultrasonic cleaner.

Drying

After the samples have been washed, they are oven-dryed to minimise metabolic activity and prepare the tissue for grinding.

Grinding and screening

The purpose of this step is to produce a uniform particle size for ease of handling and to ensure a homogeneous mixture. A mechanical grinder such as a Wiley mill is customarily used; to eliminate metal contamination that may influence micronutrient levels, all the grinding surfaces should be stainless steel. Plant tissues are ground until they are fine enough to pass through a screen with 0.5 to 1.0 mm size holes.

Determination of Chemical Composition

Destruction of organic matter

The first step in the chemical analysis of plant tissue is the destruction of the organic matter component by removing the carbon, hydrogen, and oxygen through volatilisation to produce a solution of inorganic mineral ions. There are two methods for the destruction of organic matter: dry oxidation (ashing) and wet oxidation (digestion) and both methods have advantages and disadvantages.

Dry ashing consists of oxidising the plant tissue at temperatures from 450 to 550°C for about 16 hours but has two disadvantages: possible volatilisation losses, and the chemical tie-up of certain micronutrients such as zinc, manganese, and copper which can form insoluble complexes with silica. Wet oxidation involves the use of several dangerous acids such as perchloric acid, which can cause explosions, sulfuric and nitric acid but is rapid and less prone to volatilisation or retention losses. Boron can be determined only by dry ashing because this nutrient is volatilised during wet ashing.

Wet chemistry analysis

Wet chemistry consists of traditional laboratory procedures such as gravimetric or titrimetric analysis (Table 6). The Kjeldahl method is a long standing technique that is still the most common way to measure total N in plant tissue and consists of two steps. First, the tissue sample is acid digested to convert the nitrogen to ammonium, and second, the ammonium is distilled and recovered as ammonia which is measured by sulfuric acid titration. Sulphur analysis is particularly difficult because of the problems of converting all organic forms of S to a single inorganic form and because of possible volatilisation losses during ashing.

Table 6. Common analytical methods for measuring mineral nutrient elements in tree seeling tissue

	Spectrophotometry		
Macronutrients	AAS/FES	Colorimetry	Wet chemistry
Nitrogen (N)	X	Kjeldahl	
Phosphorus (P)	X		
Potassium (K)	X		
Calcium (Ca)	X		
Magnesium (Mg)	X		
Sulfur (S)	X		Gravimetric or Titrimetric
Micronutrients			
Iron (Fe)	X		
Manganese (Mn)	X		
Copper (Cu)			
Zinc (Zn)	X		
Boron (B)	X		
Molybdenum	(Mo)	X	
Chloride (Cl)			Titrimetric

In a comparison of three methods of sulfur analysis, an autoanalyser method involving titration was found to be best for tree foliage.

Spectrophotometry

As the name implies, spectrophotometry (or spectrometry) includes several chemical analysis techniques which use specific wavelengths of electromagnetic radiation, including visible light, to quantify nutrient element concentrations in solutions prepared from plant tissue samples.

Colorimetry—Also known as visible light spectrophotometry, several plant nutrients are measured colorimetrically, especially P and B (Table 6). As an example of this technique, orthophosphate forms a yellowcolored complex when mixed with certain reagents and shows optimal absorption at a wavelength of 440 nm. The color is allowed to develop for 30 minutes and the intensity of the color is measured with a spectrophotometer. The P concentration is determined by comparison with a series of standard solutions of known P concentrations.

Flame photometry — The most common types of spectrophotometry used in plant analysis labs are flame emission spectrophotometry (FES) and atomic absorption spectrophotometry (AAS).

FES is the older of the two techniques, having been used as an analytical tool for over 100 years. The basic procedure consists of vaporising the sample by spraying it into a flame where the intense heat energy excites the atoms in the solution, causing them to emit light. The emitted light of each element is composed of certain wavelengths which are characteristic for that element. Light of the wavelength of interest is separated from the rest of the emission by a selective filter called a monochromator. A photomultiplier measures the intensity of the light which corresponds to the concentration of the element in the sample solution.

AAS is a more recent technique that was developed in the 1950's and the principle is essentially the inverse of FES. Atoms are capable of absorbing light at characteristic wavelengths and so the AAS technique consists of quantifying the light absorbed by the nutrient atoms in a solution prepared from plant tissue. The sample solution is sprayed into a flame of relatively low temperature that dissociates the elements into a gaseous vapor. This atomic vapor lies directly between a light source that emits only the spectrum of the element of interest and a photomultiplier. The atoms in the vapor absorb a portion of the light of the specific wavelength and the degree of light absorption is measured by the photomultiplier. The concentration of the nutrient element is calculated by comparison with a set of standard solutions. Any element can be measured by AAS if its light is in the spectral range of the instrument, most of which are sensitive down to 190 run. For that reason, the plant nutrients C, H, 0, N, S, and P cannot be measured by AAS since their characteristic wavelengths are below 200 nm. Other newer spectrophotometric techniques include the ICAP spectrometer which is used by a few testing laboratories. This machine is computer-equipped and can measure up to 61 different elements at one time but is also relatively expensive.

Interpretation of Seedling Nutrient Analysis

While most nursery managers do not need to be overly concerned about laboratory methodology, the interpretation of lab results is an entirely different matter. Most of us have struggled over lab reports of seedling nutrient analyses and attempted to make some sense out of them by comparing the reported nutrient values to ranges of values published in some nursery manual. The interpretation of seedling nutrient analyses requires an overall comprehension of the variation that can be expected and an understanding of how to compare the results to published standards. Skill in interpretation is only acquired through practice and experience and so professional help should be sought when considering nutrient analysis for the first time.

Units

The first thing to check when receiving a lab report is to

look at the reporting units. The majority of analytical labs report their results in concentration units although nutrient content is sometimes reported in research studies.

Concentration

Plant nutrient levels are traditionally reported in proportional units of tissue dry weight: macronutrients in percent (%) and micronutrients in parts per million (ppm). Proportional units describe how concentrated the nutrients are in the tissue. Another unit of nutrient concentration is micrograms per gram (ug/g) which is the same as ppm.

Conversion between % and ppm is sometimes necessary and is very simple:

— To convert % to ppm - multiply by 10,000

— To convert ppm to % - divide by 10,000

Content

Another unit that is sometimes seen in the research literature is related to the total amount of nutrient that a seedling or seedling component contains and is usually expressed in weight per plant (mg/seedling) or, when converted to a seedbed basis, weight per area (kg/ha).

Conversion between nutrient concentration and nutrient content is possible only if the oven-dry weight of the seedling is known. To determine nutrient content per seedbed it is necessary to know the seedbed density.

It is important to carefully distinguish between nutrient concentration and nutrient content during data interpretation. The terms are often confused in the literature which has confounded interpretation. Both concentration and content units have limitations: data reported in concentration units are subject to the dilution effect resulting from new growth, and data reported in content units do not specify the size of the plant.

There is no consensus as to which of the two reporting units are most useful and many seedling nutrition experts suggest that both concentration and content be reported. Nutrient concentration values are useful in terms of estimating seedling quality because all published nutrient standards are in concentration units but nutrient content may be the best predictor of seedling outplanting performance because it incorporates seedling size. One drawback of nutrient content is that total seedlings must be analysed or ovendry seedling weight must be measured so that concentration units can be converted to content.

Variation

Seedling nutrient status has been shown to vary with seedling age, species, ecotype, stock type, tissue type (foliage vs. whole seedling), weather and nursery as well as the variation that exists between different analytical labs. It is important to realise that this variation exists so that interpretations can be based on relevant comparisons.

Genetic

Seedling nutrient values have been shown to vary between species, ecotype, and even between individual trees. One nursery fertility expert has published separate nutrient standards for different ecotypes of the same species - interior and coastal Douglas-fir.

Nursery

Nursery environment may also affect the nutrient status

of tree seedlings because of differences in soil fertility, cultural practices, and climate. Nutrient values of conifer seedlings were found to vary between three British nurseries although manganese was the only nutrient showing large differences. In a study of Douglas-fir seedlings, both macro- and micronutrients were shown to vary not only between nurseries but between sections in the same nursery.

Laboratory

The problem of variation between analytical laboratories has been extensively debated.

Table 7. Statistical Comparison of tissue nutrient levels reported by six different laboratories

Mineral Nutrient	*Average Concentration (based on dry wt.)*	*Coefficient of Variation (%)*
Macro-nutrients	—%—	
N	2.66	2
P	0.13	4
K	1.36	1
Ca	0.47	14
Mg	0.16	10
Micronutrients	—ppm—	
Mn	1,500	3
Fe	139	14
Zn	15	6
Cu	7	24

Variation in reported values may be caused by different lab procedures or equipment. Two analytical procedures for measuring total N in seedling tissue were compared and the Kjeldahl method was found to produce consistently lower values than the Dumas procedure. Sending blind replicates of leaf tissue to six different laboratories for nutrient analysis resulted in nutrient values with coefficients of variation ranging from 1-24% (Table 7).

The only practical solution for lab-to-lab variation is to select a reputable lab so that seedling samples from one year to the next are analysed by the same procedure. Published results or standards should specify the analytical procedures used so that the interpreter of the data is aware of the possible implications.

Comparing analytical results to standard values

In order for seedling nutrient values to be meaningful, they must be compared to some standard values. As just discussed, these standard values should be as specific as possible and take into account seedling species, stock type, type of tissue and analytical procedure if possible. Most sources present standard nutrient values as ranges instead of discrete values to accommodate natural variation. Nutrient ratios may be as useful as absolute values and may point out nutrient imbalances within the seedling.

Nutrient standards for conifer foliage tissue are presented in Table 8. The problem with these "generic" nutrient standards is that they may not be sensitive enough to reveal significant differences. Due to luxury consumption of manganese by some conifer species, the guidelines in Table 8 are quite broad (100-5000 ppm) whereas the specific guidelines for Douglas-fir seedlings are more restrictive (390-1294).

Some nurseries are beginning to gather specific mineral nutrient values for their species and environments. Loblolly pine seedlings were collected from 33 southeastern U. S. nurseries by the Auburn Nursery Co-op and analysed at the same laboratory for seedling nutrients to provide base data for soil management decisions.

Table 8. Standard values for mineral nutrient concentrations in conifer Needle tissue

Mineral nutrient	Units (% dry wt)	Adequate Range Bareroot Seedlings	Container seedlings
N	%	1.20 to 2.00	1.30 to 3.50
P	%	0.10 to 0.20	0.20 to 0.60
K	%	0.30 to 0.80	0.70 to 2.50
Ca	%	0.20 to 0.50	0.30 to 1.00
Mg	%	0.10 to 0.15	0.10 to 0.30
S	%	0.10 to 0.20	—
Fe	ppm	50 to 100	60 to 200
Mn	ppm	100 to 5000	100 to 250
Zn	ppm	10 to 125	30 to 150
Cu	ppm	4 to 12	4 to 20
Mo	ppm	0.05 50 0.25	0.25 to 5.0
B	ppm	10 to 100	20 to 100
Cl	ppm	10 to 3000	—

Foliar nutrient levels for three conifer species were reported as tentative guidelines for one western U. S. nursery.

Seedling Nutrient Status and Outplanting Performance

Physiological factors affecting outplanting performance Outplanting performance, defined as survival and growth, is related to certain seedling physiological factors which are all affected by nutrient status.

Dormancy

Ideally, seedlings are lifted and stored while dormant but

break bud rapidly after outplanting in the field. Two studies have shown that lateseason fertilisation generally increased the speed of bud break after outplanting. Benzian et al. found a positive relationship between amount of N applied and speed of bud burst in Sitka spruce, Norway spruce, and western hemlock but reported that nitrogen retarded bud break in grand fir.

Frost Hardiness

Seedlings must not only be dormant, but also frost hardy when lifted so that they are not injured by cold temperatures during refrigerator or freezer storage or after outplanting. Research studies are contradictory on the effect of seedling nutrient status on frost hardiness. Aldhous stated that heavy late-season N applications may cause seedlings to flush prematurely and could lead to spring frost damage. High fertilisation levels consistently reduced frost hardiness of Douglas-fir seedlings when frost hardiness was measured by electrical impedance.

Thompson found that fall N fertilisation increased frost hardiness of Douglas-fir seedlings while P fertiliser had a variable effect. Timmis also studied Douglas-fir seedlings and found that nutrition had a definite effect on frost hardiness and recommended a K:N ratio of 0.6 in fertiliser programs for container seedlings. Larsen reported that frost hardiness in Douglas fir was decreased at both high and low N concentrations and that maximum hardiness occurred in tissues with 1.3 to 1.4% N.

Aronsson observed that frost damage to Scots pine (Pinus sylvestris L.) increased at N concentrations above 1.8 to 2.0%; laboratory freezing tests over a range of N levels gave lowest injury at tissue N concentrations between 1.3 and 1.8%. Although there is not enough information to prescribe specific levels of all mineral nutrients for maximum frost hardiness, the best recommendation at present would be to strive for adequate but not excessive levels of all nutrients, especially nitrogen.

Root Growth Capacity

The ability of a seedling to rapidly produce new roots after outplanting is an obvious asset, especially with bareroot seedlings that need to quickly reestablish soil-root contact. Ritchie and Dunlap stated that root growth capacity (RGC) should be related to seedling nutrient status and concluded that more studies are needed. In a factorial experiment with N and P fertiliser, there were no clear relationships between the treatments and new root growth of Douglas-fir seedlings. A positive relationship between RGC and fertilisation, however, was demonstrated in another study with 2+0 Douglas-fir seedlings.

Drought Resistance

Outplanted seedlings must be able to tolerate moisture stress until their roots can extend out and contact new sources of soil moisture. Ritchie stated that seedling nutrient status can affect internal water relations: N and K can reduce transpiration rates, perhaps through osmotic adjustments, whereas P may increase water loss. High nitrogen levels have been shown to reduce drought resistance of loblolly pine seedlings and lower field survival of lodgepole pine seedlings. In a review of the relationship of fertilisation to physiological quality of seedlings, van den Driessche concluded that drought resistance of conifer seedlings can be increased by providing an adequate, but balanced, supply of N, P, and K.

Predicting outplantinq survival

Fertilisation

The effect of nursery fertilisation on outplanting survival is mixed-about half the sources reporting a positive effect and half a neutral or negative effect. Field trials with Douglas-fir seedlings showed a positive effect of fertilisation in four cases. White spruce seedlings showed increased survival after two growing seasons for both N and P fertilisation. Wilde et al. reported a positive but statistically nonsignificant effect of nursery fertilisation on outplanting survival of jack pine.

On the other hand, two studies reported little or no positive effect of fertilisation on outplanting survival for either white spruce or loblolly pine although outplanting site differences were noted. A neutral or negative effect was reported for red pine (Pinus resinosa Ait.),-Douglas-fir, and for several conifers in Britain.

Foliar nutrients

Only two studies have tried to correlate foliar nutrient levels and outplanting survival. van den Driessche attempted to form regressions between foliar N and outplanting survival of Douglas-fir seedlings. Although a positive relationship was established, only 14% of the variation in survival could be explained by foliar N level. Working with four different species of conifer seedlings, van den Driessche formulated regressions between percent survival and foliar nutrients (N, P, K) and found only three significant relationships for coastal Douglas-fir and Sitka spruce and the most These results lead to the conclusion that, although foliar nutrient levels are related to outplanting survival, the regressions are not precise enough to yield accurate predictions.

Other factors such as timing and rate of fertiliser applications, nursery cultural treatments and outplanting site conditions probably explain this lack of a useful relationship.

Predicting Growth After Outplanting

Fertilisation

In contrast to field survival, nursery fertilisation appears to be a good predictor of seedling growth after outplanting. A definite, positive relationship between nursery fertilisation and seedling growth after outplanting is apparent from the literature with eight out of nine authors reporting positive results up to 5 years after planting Positive results were reported for a variety of species:

Douglas-fir; white spruce; Sitka spruce, Norway spruce and western hemlock; and jack pine. The majority of the fertiliser treatments involved nitrogen but one micronutrient (Cu) increased shoot growth of radiata pine when outplanted on a copperdeficient podzol. The only study that did not show a positve increased in field growth involved red pine and reported slight but nonsignificant effects.

Foliar nutrients

Several studies have attempted to use foliar nutrient levels to predict growth after outplanting. The most successful of these endeavors resulted in a precise linear regres sion (R^2=0.84) between foliar N content and third-year field height of loblolly pine. van den Driessche, computed a series of linear regressions between foliar nutrient concentrations of current height growth of four northwestern U. S. conifer seedlings. The most precise of these regressions was for Sitka spruce and had a coefficient of determination (R^2) of 0.51. The same author reported a regression between foliar N concentration and secondyear height of Douglas-fir seedlings that had rather poor precision (R^2=0.18). Height increment of Norway spruce was reported to be "closely correlated" to foliar nitrogen concentration although no regression statistics were given. Larsen, et al. constructed a series of linear regressions between seedling attributes and outplanting performance of loblolly pine; foliar N content was most closely related to height growth after outplanting with an R^2=0.38.

Foliar nutrogen appears to be a reasonable predictor of growth after outplanting and nutrient content may be a better independent variable than nutrient concentration.

Limitations

Even though the relationship between foliar N levels and shoot growth are of acceptable precision, there appears to be no way to predict if the seedling will survive, which, of course, is a necessary precedent. This poor relationship to survival probably reflects the influence of other factors such as seedling morphology, physiology, and outplanting site conditions.

No Guarantee of Vitality

Just because a seedling contains an ample supply of all 13 mineral nutrients is no assurance that the seedling is alive. A healthy seedling that was killed by freezing temperatures during storage would still contain a normal complement of all mineral nutrients.

Seedling size

The concentration of mineral nutrients in a seedling is independent of seedling size, which has a definite effect on outplanting performance. Nutrient content, on the other hand, does reflect seedling size because a larger seedling will contain more total nutrients. Hoyle and Mader studied relationships between foliar nutrients and growth of red pine and found that higher correlation coefficients were almost always related to nutrient content rather than nutrient concentration.

Seedling storage

The effect of cold storage on seedling nutrient status has not been widely studied and, because seedlings would normally be stored after nutrition samples were taken, could affect the value of nutrient levels as a predictor. Intuitively, proper storage should have no effect on seedling nutrient content since no mineral nutrients are being lost or gained. Nutrient concentration, on the other hand, could change during the storage period if temperatures were high enough to allow dry weight losses through respiration. van den Driessche found that higher nursery fertiliser levels reduced outplanting survival of Douglas-fir seedlings which were fall or winter lifted and suggested that high seedling nutrient levels could increase storage damage.

Outplanting site

Another complicating factor that could influence the precision of correlations between seedling nutrient status and outplanting performance involves environmental conditions on the planting site.

10

Recalcitrant Seeds

Generally, orthodox and recalcitrant seeds undergo a period of drying during their maturation and are shed at low water content which is in equilibrium with the prevailing relative humidity. The equilibrium water content at any particular r.h. is determined by seed composition, but all orthodox seeds can withstand dehydration to around 5 percent (0.053 g H_2O g^{-1} dry material [g g^{-1}]), even when maturation drying is not completed prior to shedding. Any seed that does not behave this way is not orthodox, and, in fact, the seeds of a great number of tropical species may accordingly be nonorthodox. Nonorthodox seeds have so far been described as either being recalcitrant or intermediate according to their storage behaviour.

Recalcitrant seeds are those that undergo little, or no, maturation drying and remain desiccation sensitive both during development and after they are shed. The situation is, however, far more complex than this because of the wide range of variability among recalcitrant seeds of different species and, indeed, of individual species under different conditions. Such seeds are shed hydrated, but the water content can generally be anywhere in the range from 0.43 to 4.0 g g^{-1}, which is 30 to 80 percent on a wet mass basis. Shedding water content is partly species characteristic, depending on the degree of dehydration that occurs late during seed development; this has been suggested to be correlated with the degree of desiccation tolerance developed by individual species.

Recalcitrant seeds are not equally desiccation sensitive, in that variable degrees of dehydration are tolerated depending on the species. This implies that the processes or mechanisms that confer desiccation tolerance are variably developed or expressed in the nonorthodox condition. As diverse mechanisms have been suggested to be involved in the acquisition of desiccation tolerance and maintenance of the integrity of dehydrated orthodox seeds, it should be appreciated that any one of these may be absent, or present but ineffective, in recalcitrant seeds. Another important consideration is that desiccation tolerance is probably controlled by the *interplay* of mechanisms or processes, *and not by any one, acting in isolation*. Thus, the absence or incomplete expression of any factor proposed to confer dehydration tolerance could have profound consequences on the ability of a seed species to withstand a measure of dehydration below a particular level of hydration.

Differential desiccation sensitivity among recalcitrant seeds of various species is clearly shown by their different responses when subjected to the same drying regime–those of some species tolerating only a slight degree of dehydration, but others surviving to far lower water contents. There are also marked differences in the rates at which water will be lost from seeds of various species under the same dehydrating conditions.

Other factors, too, influence the postharvest responses of recalcitrant seeds, e.g. developmental status and chilling sensitivity. In terms of desiccation sensitivity alone, therefore, it is not merely that a seed species *is* recalcitrant, but rather, *how recalcitrant* it is. This fact led to the proposal of a continuum of recalcitrant seed behaviour, from species that are highly desiccation –and probably also chilling–sensitive, to those that will tolerate drying to the lowest water content still commensurate with recalcitrant seed behaviour and will also tolerate relatively low temperatures.

The concept of a continuum of postharvest seed behaviour extends beyond the category of recalcitrant seeds. The continuum grades from extreme desiccation-sensitive types through the minimally recalcitrant types, to the intermediate seed species that do not react adversely to low temperatures, through those that are chilling sensitive when dehydrated, and finally to orthodox seeds that will tolerate less or more extreme dehydration. It is also possible that there are seed species that behave in a manner that characterises them as lying *between* the hitherto-defined categories: recalcitrant, intermediate, and orthodox.

The idea of an extended continuum of seed behaviour from the most desiccation tolerant of orthodox species, to the recalcitrant species that are most sensitive to even slight water loss, embodies many properties of seeds and their responses. It has its foundations in an appreciation of the physiological status of seeds at various water potentials and the properties of water at the various hydration levels corresponding to specified water potential ranges. It is more meaningful to consider seed responses to dehydration in terms of water potential rather than water content, but, since these two measures can be loosely correlated, the more familiar water content terminology is used here.

According to Vertucci and Farrant: "Discrete changes in metabolic activity with moisture content are hypothesised to be associated with discrete changes in the physical properties of water... Thus upon the loss of water with certain properties, an essential function provided by [that] water is no longer possible. A tissue that is not damaged by the removal of a certain type of water has developed mechanisms to tolerate or avoid that particular stress." While the discussion that follows is not dependent on the reader's appreciation of the differences in the types of intracellular water, the basis of the arguments presented is that sequential removal of water with specific properties will have particular damaging effects on seed tissues that are not possessed of the appropriate mechanisms or processes to counteract that damage.

Desiccation Tolerance Mechanisms

It is most expedient to consider the processes or mechanisms listed below, which might confer protection against desiccation, and their deficiency or absence, which could contribute to the *relative degrees of desiccation sensitivity*.

- Intracellular physical characteristics such as
  - reduction of the degree of vacuolation,
  - amount and nature of insoluble reserves accumulated,
  - integrity of the cytoskeleton,
  - conformation of the DNA, chromatin, and nuclear architecture.
- Intracellular de-differentiation, which effectively results in the minimisation of surface areas of membranes and probably also of the cytoskeleton.

— "'Switching off" of metabolism.
— Presence, and efficient operation of, antioxidant systems.
— Accumulation and roles of putatively protective molecules, including late embryogenic accumulating/abundant proteins (LEA's), sucrose and certain oligosaccharides, or galactosyl cyclitols.
— Deployment of certain amphipathic molecules.
— An effective peripheral oleosin layer around lipid bodies.
— The presence and operation of repair mechanisms during rehydration.

Although the interrelationships among them are far from being resolved, these processes or mechanisms are those that have been implicated to date in the acquisition and maintenance of desiccation tolerance. However, it is important to realise that additional properties that contribute to the ability of seeds to withstand extreme dehydration are likely to be elucidated.

Intracellular Physical Characteristics

Vacuolation and Reserve Deposition

In 1957, Iljin had already identified one of the major requirements of cells of desiccation-tolerant plant material: the ability to withstand mechanical stress. Vacuole volume reduction, whether by the shrinkage of the space occupied by these usually fluid-filled organelles or by their becoming filled with insoluble reserve material, is one of the mechanisms that would contribute to increased mechanical resilience of cells to dehydration. This aspect was examined by Farrant and others for (1) *Avicennia marina*, the highly recalcitrant seeds of which can withstand very little dehydration either before or after they are shed; (2) *Aesculus hippocastanum*, a temperate recalcitrant species, the seeds of which overwinter in the hydrated condition during which the necessary stratification occurs to facilitate germination the following spring; and (3) *Phaseolus vulgaris*, a typical orthodox seed that attains a low water content prior to shedding and is long-lived in this condition.

Avicennia marina seeds lose no water during development, and are as sensitive to dehydration before shedding as after abscission. These seeds, at best, are unable to surive water contents lower than 0.5 g g^{-1}. The vacuoles ultimately occupy almost 60 percent on average of the volume across the cells of all axis tissues, and 90 percent of the cotyledonary cells when mature. At no stage do either the axial or cotyledonary vacuoles contain insoluble reserves, the little insoluble reserve material occurring as plastid starch.

Seeds of *A. hippocastanum* naturally undergo a measure of dehydration during development, accompanied by an increase in relative desiccation tolerance. The mature seeds are more desiccation tolerant than those of *A. marina*, being able to withstand dehydration to water contents in the range of 0.42 to 0.25 g g^{-1}. Vacuoles ultimately constitute only a small fraction of the intracellular volume, particularly in the axis cells at maturity. The cotyledonary cells contain many large, starch-filled plastids and protein bodies and are considerably less vacuolate than those of *A. marina*.

In *P. vulgaris* seeds, which are orthodox and able to tolerate low water contents, vacuolar volume is reduced to an insignificant proportion in axis cells, and vacuoles in cotyledonary cells accumulate an amorphous, presumably insoluble, material. The differential degree of vacuolation and insoluble reserve deposition among the three species, in both developing and mature seeds, correlates with their degree of desiccation sensitivity. This is in accord with the concept that a high degree of vacuolation can lead to lethal mechanical damage upon dehydration.

Reaction of the Cytoskeleton

The cytoskeleton, the major components of which are microtubules and microfilaments, is not only an integrated intracellular support system, it also plays a major role in imposing organisation on the cytoplasm and also the nucleus. Microtubules consist of polymerised a - tubulin, and microfilaments are composed of F-actin, which is a polymer of G-actin. We are presently investigating the status of the actin microfilaments in hydrated and variously dehydrated embryonic axes of seeds of *Quercus robur*, a temperate recalcitrant species. In the hydrated state, there is an extensive microfilamentous network in the cells of the root tip, which becomes dismantled as the seeds are increasingly dehydrated–a feature that is expected for orthodox seeds as well.

In such desiccation-tolerant seeds, orderly reassembly of the elements of the cytoskeleton accompanies imbibition, but once the water content falls to damagingly low levels in *Q. robur*, the microfilaments are not reassembled when the seeds are subsequently rehydrated. The resultant lack of the intracellular support and structural organisation afforded by the cytoskeleton would obviously be a major damaging factor upon rehydration of recalcitrant seeds. Additionally, certain cytomatrical (cytoplasmic) enzyme systems exist as multienzyme particles in plant cells, the formation of which could occur because of the binding of key or anchor enzymes to the microfilaments of the cytoskeleton, as illustrated for glycolysis by Masters. Thus, failure of the cytoskeleton to reassemble following deleteriously low levels of dehydration would have physiological as well as structural consequences in the cells of desiccation-sensitive seed tissues.

DNA, Chromatin, and Nuclear Architecture Conformation

Maintenance of the integrity of the genetic DNA material in the desiccated condition in orthodox seeds, and/or its rapid repair when seeds are rehydrated, is considered to be a fundamental requirement for desiccation tolerance. There is, however, little information on which to draw. DNA assumes different conformational states depending on water activity and, although this has not yet been demonstrated for seeds, it is considered that as water is lost (i.e. water activity is lowered) such conformational changes will occur.

There is an increase in the number of base pairs per turn of the DNA helix as water is lost from the individually hydrated phosphate groups, and water bridges are formed instead as the conformation changes from the B to the Z form. Osborne and Boubriak have suggested that protein glycation (i.e. the nonenzyinic addition of reducing sugars to [*i.a.*] histone proteins) is likely to occur, which could increase the incidence of DNA conformations appropriate to the dehydrated state.

However, besides the postulated necessity of conformational changes in the DNA occurring as desiccation-tolerant material is dehydrated, the structure of the chromatin itself must also be stabilised. The highly condensed state of the chromatin in dry, orthodox seeds, which is reversed at the stage in germination when desiccation sensitivity ensues, is thought to be a visible manifestation of its stabilised condition. A major factor in chromatin stabilisation in the dry state in orthodox seeds might be the change in the H1-histone: nucleosome ratio to 2:1 from the 1:1 ratio that typifies the hydrated condition.

Nuclear architecture is a further factor that is probably involved in chromatin stability. The structural framework of the nucleus has been convincingly demonstrated for plant cells and is based on intermediate-type filaments called lamins. The nucleoskeleton, organised into the lamina (underlying and connected to the inner surface of the nuclear envelope) and matrix (ramifying throughout the nucleus) is suggested to support and localise the chromatin in discrete domains, imposing the topological organisation and coordination of intranuclear processes. It is implicit that during dehydration and in the desiccated state of orthodox seeds, orderly reorganisation of the nucleoskeleton should occur with its restitution as a functional framework upon rehydration.

While little is known about the effects of dehydration on the DNA, chromatin, and nuclear architecture in desiccationsensitive seeds, their stability in the dehydrated state clearly must be a prerequisite for desiccation tolerance. Maintenance of the integrity of the nucleus as a whole, and the genome in particular, may be imperfectly expressed, or the ability for this may even be totally lacking, in recalcitrant seeds.. What is equally likely is that DNA repair mechanisms themselves are inadequate to restitute damage caused by dehydration of desiccation-sensitive seeds.

Intracellular De-differentiation

De-differentiation, a characteristic of maturing desiccationtolerant seeds, is essentially a means by which intracellular structures are simplified and minimised, which strongly suggests that membranes and cytoskeletal elements are vulnerable to dehydration. This phenomenon is reversed in orthodox seeds when water is taken up early during germination.

An examination of the quantitative and qualitative status of mitochondria in seeds of *Avicennia marina*, *Aesculus hippocastanum*, and *Phaseolus vulgaris* showed that the proportion of cell volume occupied by these organelles was highest in *A. marina*, which is very desiccation sensitive, and substantially less for *A. hippocastanum*, which is in keeping with its less recalcitrant nature. In *P. vulgaris*, mitochondria occupied a significantly smaller proportion of the cell volume, even preceding the onset of maturation drying. Also, the mitochondria occupied a far greater proportion of the cell volume in the axis meristems of the two recalcitrant species than in the orthodox species, *P. vulgaris*.

There were also marked differences in the structural complexity of the mitochondria among these three species: *A. marina* and *A. hippocastanum*, had well-developed cristae and a structure that was generally typical of an active, hydrated plant tissue; while in *P. vulgaris*, the mitochondria were almost completely de-differentiated even at tissue water contents comparable to those of the recalcitrant species at shedding. It thus seems that retention of organelles in the highly differentiated state is a major factor in the desiccation sensitivity of recalcitrant species, whereas the ability for ordered de-differentiation is, in fact, a prerequisite for seed survival in the dehydrated state.

There has long been uncertainty as to whether dehydration *causes* de-differentiation, or this intracellular minimisation actually *precedes* the initiation of maturation drying. However, the observations on *P. vulgaris* reported by Farrant and others, indicating that mitochondrial de-differentiation occurs, and that respiratory rate declines markedly before maturation drying, support the idea that substantial qualitative and quantitative change actually occurs in advance of water loss.

"Switching off" of Metabolism

Electron transport, albeit at a low level, has been recorded for dehydrated plant tissues, and respiration is measurable even at seed water contents as low as 0.25 g g^{-1} [20 percent, wmb]. However, in the water content range 0.45 to 0.25 g g^{-1}, unbalanced metabolism may lead to the generation, and essentially uncontrolled activity, of free radicals. It is therefore imperative that, during maturation drying, desiccation-tolerant seeds be able to pass through this water content range with the minimum of damage. The efficient operation of antioxidant systems, as well as the "switching off" of metabolism, would reduce such damage.

Rogerson and Matthews recorded that a sharp decline in respiratory substrates precedes, and presumably causes, the fall in respiratory rate which, they suggested, is an essential event enabling an orthodox seed to withstand rapid loss of water. The observations of Farrant and others, indicating that a decline in respiratory rate occurs while mitochondria become substantially de-differentiated *prior* to maturation drying in the orthodox seeds of *Phaseolus vulgaris*, support the data and suggestions of Rogerson and Matthews. In desiccation-sensitive seeds, lethal damage occurs in the water content range 0.45 to 0.25 g g^{1} and, in some species, at considerably higher levels.

Death of relatively hydrated recalcitrant seeds occurs when water is lost slowly. However, rapid dehydration rates allow survival to lower water contents. This observation led initially to the use of relatively rapid air-drying of excised embryonic axes to facilitate cryostorage and later to the development of the flash-drying technique, by which the axes are dehydrated much more rapidly.

Flash-dried axes are *not* desiccation tolerant; on the contrary, they will not survive for longer than a day or two at best, under ambient conditions although they may be cryostored successfully. The desiccation sensitivity of recalcitrant material is the outcome of the fact that the axes (seeds) are actively metabolic, and the success of very rapid dehydration is that it minimises the effects of this metabolism. Damage occurring in conjunction with unbalanced metabolism at these relatively high water contents should not be confused with desiccation damage in the strict sense. The latter describes the damage that occurs when water that is required to maintain the integrity of intracellular structures is removed.

Desiccation damage *sensu stricto* is the consequence of removing (any, or some, depending on the species) structure-bound, nonfreezable water. Lethal damage occurs upon loss of this water, even if flash-drying has successfully maintained axis viability to, or close to, this level of hydration. Another critical aspect of ongoing metabolism is cell cycling. The cell cycle describes the nuclear DNA content as 2C in cells that are not preparing for nuclear division, and as 4C in cells in which DNA replication has occurred, where the constant, C, denotes the DNA content of the haploid condition.

During the cell cycle four distinct phases can be identified, *viz.* the G_1 phase (2C), which is followed by the S phase, during which DNA replication occurs; after this the cells enter the G_2 phase, during which the amount of DNA remains doubled (i.e. 4C) as a result of events in the S phase, and this is followed by the phase known as G_2M, when mitosis reduces the DNA content to the 2C level typical of somatic cells in the next G_1 phase. Brunori found that in orthodox *Vicia faba* seeds, most of the cells were arrested in G_1, and that DNA replication was one of the first events to be curtailed as the embryo cells lost water. S-phase replication is resumed only after several hours of imbibition, when water again becomes available to postharvest, orthodox seeds, as shown by Sen and Osborne for *Secale cereale* (rye): as soon as replication to 4C values occurs and the cells enter G_2M, desiccation tolerance is lost.

In the the highly recalcitrant seeds of *Avicennia marina*, there is only the most transient arrest of DNA replication in root primordia (meristems) of *Avicennia marina* lasting no more than 24 hours around shedding. This is the time when the seeds (although highly desiccation sensitive) are *relatively* most tolerant of water loss and least active. Ongoing cell cycling is associated with marked desiccation sensitivity of the DNA. When only 16 to 18 percent of the total water is lost from the *A. marina* material, there is a reduction of 70 to 80 percent in the nuclei that will incorporate thymidine, and after a 22-percent water loss, damage of the DNA cannot be repaired even when water is made freely available.

Ongoing cell cycling, therefore, is another manifestation of the fact that metabolism is not "switched off," at least in these highly recalcitrant seeds, which is considered to be a major factor accounting for their desiccation sensitivity. In related work on the temperate recalcitrant species *Acer pseudoplatanus*, however, cell cycling was found to be arrested, with over 60 percent of the cells in the 2C state. However, seeds of *A. marina* are poised for immediate germination, while those of *A.*

pseudoplatanus are dormant, requiring cold stratification before they will germinate. For seeds of *Azadirachta indica*, recorded as showing intermediate behav-ior, the 2C DNA level has been reported as occurring to the virtual exclusion of 4C. These disparate results on the status of the cell cycle in three nonorthodox seed species serve to highlight the fact that different factors may contribute to the nature, and differing degrees, of desiccation sensitivity.

Operation of Antioxidant Systems

A range of antioxidant processes operate in orthodox seeds, and the role of such processes under conditions of water deficit and desiccation stress in plants, It is particularly in the water content range from 0.45 to 0.25 g g^{-1} (30 to 20 percent, wmb), that unregulated metabolic events resulting in the first wave of free-radical generation are likely to occur. This implies that antioxidant systems (i.e. free-radical scavenging systems) should be maximally effective during maturation drying of orthodox seeds, and again when seeds take up water upon imbibition.

Reviews of metabolic damage associated with dehydration of recalcitrant seeds highlight the idea that free-radical generation may well be a major injurious factor, particularly because protective mechanisms appear to become impaired under conditions of water stress. Rapid formation of free radicals and decreasing activity of antioxidant systems have been reported as occurring during dehydration of the seeds of the temperate recalcitrant species *Quercus robur*.

Lipid peroxidation, which is a major consequence of uncontrolled free-radical generation, with the ultimate accumulation of a stable free radical in the embryonic axes, has been shown to accompany dehydration of the seeds of three temperate, recalcitrant species— -*Q. robur*, *Castanea sativa*, and *Aesculus hippocastanum* –and free radical formation has been reported to accompany viability loss in seeds of the highly recalcitrant, tropical species *Shorea robusta*. While hydroperoxide formation has been shown to accompany dehydration at a range of temperatures of the recalcitrant seeds of *Zizania palustris*, significantly more was produced at 37 °C than at 25 °C, and tetrazolium tests revealed that viability was severely affected by water loss at the higher temperature.

There is no doubt that damage ascribable to uncontrolled free-radical generation occurs during dehydration in the recalcitrant seeds of a range of species that show differing degrees and manifestations of nonorthodox behaviour. This implies not only that free radicals are produced as a consequence of water stress in these desiccation-sensitive seeds, but also that antioxidant systems are ineffective at curbing them. Together, then, these factors must be seriously considered as constituting one of the major causes of desiccation sensitivity.

Putatively Protective Molecules

Late Embryogenic Accumulating/Abundant Proteins (LEA's)

LEA's comprise a set of hydrophilic, heat-resistant proteins associated with the acquisition of desiccation tolerance in developing orthodox seeds. Their synthesis appears to be associated with the high ABA levels that peak during the later stages of seed development. The characteristics of LEA's and the conditions under which they appear have led to suggestions that they function as protectants, perhaps stabilising subcellular structures in the desiccated condition.

The position of LEA's (or dehydrin-like proteins, as they may be termed) in nonorthodox seeds appears at first sight to be anomalous, as some species do not express these proteins while others express them to variable extents. Seeds of *Avicennia marina*, which are extremely desiccation sensitive, appear not to express LEA's at all. In contrast, seeds of *Zizania palustris* (North American wild rice), which are recalcitrant but show differential responses to dehydration depending on temperature, do express this type of protein.

Dehydrin-like proteins were shown to be expressed in a range of temperate, recalcitrant species, but the absence of such proteins correlated with low ABA levels was found to characterise the mature, recalcitrant seeds of 10 tropical, wetland species. Those authors showed the presence of dehydrin-like proteins in other temperate and tropical recalcitrant species, and suggested that their occurrence may be habitat-related, perhaps also providing protection against low-temperature stress. In a comparative study on mature seeds of two tropical tree species, neither of which occurs in wetlands, dehydrin-type proteins were absent in *Trichilia dregeana*, while accumulating in *Castanospermum australe*. The immature seeds and the seedlings of these two species were shown to differ in terms of production of such proteins in response to stresses imposed by dehydration, application of ABA, or exposure to cold, with *T. dregeana* not responding by the production of these putatively protective proteins.

Thus, it seems that the ability to express LEA's or dehydrin-type proteins cannot be taken as an indication that the seeds of a particular species will or will not withstand dehydration. This indicates clearly that desiccation tolerance must be the outcome of the interplay of more than one (and probably many) mechanisms or processes. Details of this, particularly pertaining to LEA's/dehydrins, sugars, and various stresses, have been reviewed by Kermode. However, the variable expression of LEA's/dehydrins in recalcitrant seeds on a species basis may, in association with the presence or absence of other factors, account for the *degree* of nonorthodox behaviour exhibited under a particular set of circumstances.

Sucrose, Oligosaccharides, or Galactosyl Cyclitols

The possible role(s) of nonreducing sugars in relation to desiccation tolerance in seeds has been extensively reviewed. Accumulation of nonreducing sugars, particularly of the raffinose series and/or galactosyl cyclitols has been implicated in the acquisition and maintenance of the desiccated state in orthodox seeds, generally in two major ways. These are in terms of the "Water Replacement Hypothesis" and vitrification, otherwise referred to as glassy state formation. Orthodox seed maturation invariably seems to be accompanied by the accumulation of nonreducing oligosaccharides which coincides with the reduction of monosaccharides, and maintenance of the desiccated state is associated with high levels of sucrose and other oligosaccharides.

Evidence for the replacement of membrane-associated water is equivocal, and a recent critique questions its relevance in the desiccated state of orthodox seeds. However, the role of sucrose in the formation of intracellular glasses (i.e. vitrifica-tion) is more convincing. The metastable, glassy state occurs *at low water contents* in seeds, when sucrose and certain oligosaccharides or galactosyl cyclitols form high-viscosity, amorphous, super-saturated solutions. The occurrence of glasses is held to impose a stasis on intracellular reactivity, protecting macromolecules against denaturation and possibly preventing or minimising liquid crystalline gel phase transformations of the lipid bilayer of membranes.

Walters and others have suggested that a significant proportion of the sugars may be tightly associated with LEA's–these complexes acting to control and optimise the rate of water loss during dehydration of orthodox seeds. It should be noted, however, that this should not obviate the participation of either the LEA's or the sugars in the maintenance of orthodox seed viability in the desiccated state. The formation of intracellular oligosaccharides occurs at the expense of monosaccharides, and confers the advantage that immediately available respiratory substrates are removed. This would serve to reduce the spectrum of damaging reactions that can occur as orthodox seeds pass through critical water content ranges favouring unbalanced metabolism, during maturation drying.

Whatever the role(s) of sucrose and oligosaccharides or galactosyl cyclitols may be in orthodox seeds, *seeking parallels* for desiccation-sensitive seeds is entirely inappropriate. While sucrose and other oligosaccharides are produced in some of the few recalcitrant seed species that have been assayed, glass formation will occur only at water contents *well below* the lethal limit. When recalcitrant seeds are dehydrated under ambient conditions (which is what would occur in the natural habitat), they lose viability at relatively high water contents–in the region of 0.7 g (or more) water per g dry mass [40 percent, wmb], which are *far* higher that those required for glass formation to occur. The same argument holds if water replacement by sugars is an operative phenomenon in orthodox seeds; this too would occur only at water contents of 0.3 g per g dry material, which is well below the lethal limit for slowly drying recalcitrant seeds.

The one involvement of sugars in the variable desiccation sensitivity of recalcitrant seeds might be via the mechanism suggested by Walters and others for maturing orthodox seeds, *viz.* the modulating effect of sugar/LEA complexes on dehydration rate. Very marked variability occurs in the rate at which recalcitrant seeds of different species lose water under the same conditions and it is possible that the significance of sugars and LEA's in embryos of recalcitrant seeds of some species lies in the modulation of the drying rate by complex formation. Walters and others have also suggested that LEA proteins in temperate recalcitrant seeds may play a role in their survival during overwintering.

Development of Amphipathic Molecules

It has been suggested that partitioning of endogenous amphipathic molecules (amphipaths) into membranes upon water loss may be a prerequisite for desiccation tolerance. Few evidence of the movement during dehydration of both introduced, apolar spin probes and endogenous amphipaths into the bilayer of desiccation-tolerant pollen. This process, which was complete after dehydration to the relatively high water content of 0.6 g per g dry mass (37 percent, wmb), was reversed during rehydration, when the amphipaths repartitioned to the cytomatrix. This reverse movement was suggested to account for the transient leakage that is invariably observed when desiccation-tolerant material is imbibed from the dry state.

The partitioning of amphipathic molecules into the bilayer was suggested by those authors as serving to maintain the integrity of membranes in the dry state in desiccation-tolerant organisms, by substantially lowering the water content at which the phase change of membrane lipids occurs. Liquid crystalline to gel phase changes in membranes are well documented in response to dehydration, but the essential property for desiccation tolerance is that they must be reversible, reestablishing the membranes in a functional condition upon rehydration. This demands that integral membrane proteins retain their position in the desiccated state, a role that might also be ascribed to the amphipathic molecules.

If the partitioning of amphipaths into membranes is established as a universal phenomenon occurring during dehydration of orthodox seeds, it is possible that they are absent or, if present, incompletely functional or nonfunctional in desiccation-sensitive seeds. Dehydration of the embryos from recalcitrant *Camellia sinensis* seeds was found to induce a phase change in membrane lipids, which was reversible, but the proteins were irreversibly affected. It may be significant that at a water content of 0.6 g g^{-1}, when amphipath partitioning has been observed to be complete, slowly dried recalcitrant seeds, and even the flash-dried axes of certain species, will have lost viability.

In highly desiccation-sensitive recalcitrant seeds, it is possible that phase changes of the membrane bilayers might not be reversible, for example, if nonbilayer structures or hexagonal phases result. Partitioning of endogenous amphipaths into the bilayer upon dehydration is unlikely to act in isolation; thus, even if such molecules are present in cells of recalcitrant seeds, they may well depend on another mechanism or process to achieve their reversible migration.

Oleosins

The term *oleosin* refers to a unique protein type that surrounds the lipid (oil) droplets in plant cells. Oleosins have a central, hydrophobic domain that interacts with the periphery of the lipid, and an amphipathic N-terminal domain that, with the C-terminal domain, facilitates interaction with the aqueous cytomatrix. The oleosin boundary of lipid bodies allows these hydrophobic masses to be accommodated as discrete entities in the aqueous cytomatrix under hydrated conditions, and it has been suggested that their role during dehydration prevents the bodies from coalescing in desiccation-tolerant seeds.

Leprince and others recorded a lack of oleosins in desiccation-sensitive seeds of some species, and although little obvious change in the integrity of the bodies as a consequence of dehydration was observed, rehydration appeared to have deleterious effects on their stability. Coalescence of lipid bodies is a common abnormality accompanying deterioration, even in cells of stored, orthodox seeds. Although the effects of fungi associated with both stored orthodox and recalcitrant seeds in bringing about lipid body coalescence cannot be ruled out, the occurrence of this phenomenon could well be, at least partly, a consequence of some deficiency in desiccation-sensitive seeds.

In view of the findings of Leprince and others, the deficiency of an adequate oleosin sheath around the lipid bodies may underlie the inherent instability of these organelles during rehydration following damaging levels of desiccation of some recalcitrant seeds. However, it must be stressed that the presence of fully functional oleosins cannot, in itself, account for desiccation tolerance. Rather, it must be viewed as one of the mechanisms contributing to the spectrum of properties necessary if orthodox seeds are to survive extreme dehydration.

Operation of Reapir Mechanisms in Rehydration

There is both indirect and direct evidence that repair mechanisms do come into play when dry orthodox seeds are rehydrated. For example, seeds that have been stored under adverse conditions, but are still 100-percent viable, typically show a lag before there are visible signs of germination, during which it is commonly accepted that repair processes are taking place. Ultrastructural studies on maize seeds have provided evidence supporting this contention, where mitochondrial repair was observed during the lag period. Studies on rye seeds have shown that even in the dry state there is progressive deterioration of the DNA as a result of endo- and exonuclease activity during storage, which cannot be repaired until the seeds are rehydrated.

Much of the evidence for the operation of repair processes during rehydration comes from osmopriming experiments on low-vigor seeds. This process involves controlled rehydration to the end of phase II, which achieves a hydration level that facilitates repair but precludes germination proper. Those authors have shown that replacement of damaged rRNA occurs, and lesions in the DNA and protein-synthesizing systems are repaired, during priming. It is generally agreed that free radical generation continues in air-dried orthodox seeds during storage and the ensuing damage obviously must be repaired on rehydration, arguing strongly for the presence and efficient operation of antioxidant systems at this stage. During dehydration of desiccation-sensitive seeds and seedlings, however, such systems have been shown to fail and are assumed to remain ineffective when water is once again provided.

When recalcitrant seeds or axes excised from such seeds are subjected to nonlethal dehydration, it is generally observed that there is an increase in the time taken for the onward growth of germination, which might be interpreted as facilitating repair. However, this is likely to be strictly limited; present studies have shown that after 22 percent of the water is lost from hypocotyl tips of *Avicennia marina*, dehydration-associated DNA damage can no longer be repaired when water is once again provided. DNA instability to dehydration is also shown by seedlings produced from orthodox seeds, once they have reached the stage when desiccation tolerance has been lost.

Very little work that targets the aspect of possible repair of mature, dehydration-damaged recalcitrant seeds has yet been done. It is presently tacitly assumed that the necessary repair systems are present, but are themselves damaged by dehydration beyond certain limits–limits that might vary among seed species of markedly differing desiccation sensitivity. However, this aspect requires considerable investigation to obtain both qualitative and quantitative data to clarify the situation.

Drying Rate

We now know that much confusion has occurred in comparative work on individual species of recalcitrant seeds because of conflicting data regarding "critical water contents," below which viability will be lost. This is because the dimensions of the *time* taken for water to be lost, or the temperature at which the drying experiments were carried out, have been ignored. While the effects of temperature will not presently be discussed, there are

axis water contents that signs of deterioration were noted, which coincided with declining viability.

At no stage did the extensive degradation that characterised axis cells from slowly dried seeds occur, supporting the proposal that if desiccation-sensitive material can pass quickly enough through water content ranges at which lethal reactions are prevalent, then it is possible to dry the material down to a far lower hydration level. There will be a water content at which *rapidly dried* material that is desiccation sensitive will sustain injury, and, while the value varies from species to species, it is usually near the range where only structure-associated (nonfreezable) water remains. Damage occurring at such relatively low water contents is defined as desiccation damage in the strict sense and is suggested to coincide with the perturbation of the nonfreezable water. In contrast, desiccationtolerant material can withstand the removal of a considerable proportion of this water.

Slowly dried desiccation-sensitive material sustains damage at relatively high water contents, certainly those where solution (i.e. freezable) water prevails. This damage is suggested to result from aqueous-based, degradative reactions that are the result of unbalanced metabolism. Recalcitrant seeds (and, indeed, probably all nonorthodox types) are hydrated and metabolically active when shed. As water is slowly lost, metabolism will continue, but when the seeds are still at relatively high water contents, metabolism will become unbalanced or out-ofphase as a result of internal water stresses. A likely consequence of this unregulated metabolism will be the generation of free radicals and accompanying oxidative damage. The severity of this type of damage, which is being termed metabolic damage, is predicted to increase in inverse proportion to the drying rate, with viability loss occurring at increasingly high water contents.

11

Vegetable Seed Quality

Some vegetables are harvested for market-use such as carrot, onion, radish, brassicas, and lettuce are typically grown in areas with lower temperatures from those required for a seed crop in order to maintain good vegetative growth and minimise bolting. Seed production of these crops often requires higher temperatures to encourage bolting although extremely high temperatures also hinder pollen growth and fertilisation which lead to reduced seed number per floret or fruit. This is true in onion, tomato, and lettuce. In contrast, beet and carrot market-use crops have been selected to avoid bolting in mild climates but they still require adequate periods of cool temperatures (vernalisation) for flowering and seed production.

The Pacific Northwest of the United States provides such sustained cool temperatures and this is the reason why beet and carrot seed production occurs in this region while market-use production of these crops is geographically more diverse. The requirement of some vegetables for specific photoperiods before induction of flowers also limits those areas where vegetable seed production is successful. This effect is dependent on the latitude and the length of the photoperiod which greatly affect seed yield. For example, onion is a vegetable crop that exhibits a wide diversity in photoperiod response and this influences where seed production occurs.

Seed production of long-day onions occurs at latitudes greater than 45o N and short-day onions at latitudes no greater than 35°N. Limited seed production also can be accomplished in glasshouses where supplemental light is provided or light excluded to promote flowering. Even when appropriate photoperiods are present, temperatures must be suitable for the completion of the entire life cycle of the crop during the growing season. Successful market-use crops such as onion, radish, broccoli, and cabbage are harvested before flowering occurs. As a result, these crops are grown in regions with cooler temperatures not suitable for seed crops which require longer periods and warmer temperatures for maturation of seed.

Precipitation amount and distribution also varies for market-use and seed crops. Many vegetable crops harvested for their vegetative structure require large amounts of early season rainfall for most rapid development. Seed crops, in contrast, demand uniform precipitation during development followed by dry conditions during seed maturation. Under these environments, dependable conditions for seed harvest are possible and pathogen infestations minimised. This is the principal reason much of the vegetable seed production has shifted to the western United States.

The Mediterranean climate of the Pacific Northwest assures cool, dry summers and warm, wet winters for optimum development of many biennial seed crops. Southern California and much of Arizona are hot and dry during the summer which is ideal for vegetable seed production. The dry condition of these locations permits seed crops to be irrigated with the proper amount of moisture at the proper time so that optimum seed quality is achieved.

Vegitable Seed Production

The following represents general requirements and considerations utilised in vegetable seed production.

Tillage

The preparation of a soil site for vegetable seed production differs little from that for fresh market production. Soils should have a high water holding capacity to encourage uniform vegetative growth. Depending on the crop and water availability, planting is done on raised, shaped, or flat planting surfaces which require appropriate soil preparation. In all cases, the seed bed must be uniform because most vegetable seed crops are precision planted and uniform emergence and seedling development necessary for optimum performance.

Planting

The planting of vegetable seeds for seed production is generally done by direct drilling or transplanting of glass house grown seedlings. Biennial vegetable seed production, however, can be achieved using two planting techniques. The "seed-to-seed" method is done by planting biennial seed, allowing the resulting plant to overwinter without being transplanted, and harvesting the seed crop the following season. This method does not permit the selection or roguing of root or other genotypic characters. A majority of carrot and onion seed crops are produced by this method which is less expensive and complex than the "root-to-seed" method.

The "root-to-seed" method is done by planting biennial seed, removing the resulting plant from the soil (in some cases stored), replanting, and harvesting the seed crop the following season. The replanted plant is referred to as a "steckling." This method permits the selection or roguing of root or other genotypic characters. It also decreases the time required for varietal development. For example, onion bulbs adapted to dry regions can be planted in the spring in the desert regions of the southwest and then shipped for seed production to central California or the Pacific Northwest for planting in the early fall of the same year. This is the most expensive method of vegetable seed production due to transportation, handling, and replanting costs, but the benefits of checking for minimum genetic quality standards before replanting are important in vegetable seed production.

Vegetable seed crops must be planted so that sufficient vegetative development occurs to support optimum fruit and seed development. In addition, the increasing emphasis on hybrid seed production necessitates planting of male and female lines at times that assure successful pollination, a process known as nicking. This may require that genetically different male and female lines be planted at differing times so that synchronous flowering occurs. Most hybrid broccoli is transplanted and staggered plantings are necessary for the male and female lines to ensure proper nicking. Hybrid cauliflower and squash are direct seeded but because of the differences in time required for the male and female lines to flower, they are planted in a relay fashion where the earlier flowering line is planted first followed by the later flowering line.

Vegetable seed production also is unique from most agronomic crops because vegetables such as beets, carrots, and cabbage are biennials and must develop sufficient vegetative growth prior to cool temperature exposure in order that vernalisation successfully induces flower formation the following season. Planting such crops too early causes winter kill or late season pest infestations. Planting too late results in a lack of vernalisation which limits flowering and reduces seed yield. In some instances, vegetable seed producers opt not to start the seed crop by seed. Rather, to ensure optimum vegetative growth for vernalisation, they plant some developed portion of the vegetative structure. This is commonly done with carrot and beet stecklings and onion bulbs which allows for maximum production of plant biomass before vernalisation.

Vegetable seeds also vary greatly in their tolerance to soil temperatures at planting. Pea, radish, and spinach require cool soil temperatures for optimum seedling emergence. Beets, cabbage, carrot, and onion are tolerant of cool soil temperatures although they do better in warmer soils. Squash and melons require warm soil temperatures for optimum seedling emergence. In regions with high summer temperatures, vegetables such as tomato, pepper, eggplant, and cucurbits are planted in the early spring to optimise flowering and seed set in early summer. Later plantings of these seed crops are possible by transplanting young plants previously started in glass houses. This avoids the delayed establishment of plants from direct seedings due to cooler night temperatures and wetter soils encountered in the early spring.

Row spacings and planting densities of vegetable seed crops differ from those for fresh market production. Sufficient space for flower development, air movement to reduce pathogens, unrestricted access to inflorescences by pollinators, mechanical cultivation, and harvest operations are necessary for seed crops. In some cases (e.g., eggplant, pepper, tomato, muskmelon, watermelon), row spacings used for seed production are the same as those for fresh market production. In most cases, row spacings are different for vegetable seed production.

In lettuce, the size of the head will be larger than the fresh market head in order that bolting is promoted. Large-head types are planted 25 to 30 cm apart while small-head and loose-leaf types are planted 15 to 20 cm apart. Similar considerations are encountered for cabbage. Broccoli and cauliflower produce sizable seed stalks requiring rows 40 to 60 cm apart. Planting densities are often different as well in vegetable seed crops from fresh market vegetable crops. Beet seed production is accomplished at higher planting densities than beets harvested for their roots. Carrot, celery, and parsnip seed production is planted at densities much less than for root or stalk production. Higher planting densities of onions grown by the bulb-to-seed method reduces lodging of the large seed stalk scapes.

The method (wind or insect) of pollination affects row spacings and planting densities. Vegetables that are wind pollinated include sweet corn and spinach. Insect pollinated vegetables include most of the cole crops, carrot, and onion. Row spacings and planting densities are also important considerations in the production of vegetable seed hybrids. In order to maximise the amount of hybrid seed produced, the optimum ratio of female to male rows must be determined.

Fertilisation

Vegetable seed crops have similar nutrient requirements as fresh market crops. Where possible, nutrients that stimulate reproductive development can be supplemented with normal fertility regimes. Boron, for example, increases the number of fruiting sites and seed produced per pod in legumes such as pea and bean. Molybdenum also is essential for optimum nitrogen fixation of legume seed crops. Another approach to stimulating reproductive development in vegetable seed crops is to split the application of fertilisers at planting and before flowering. This avoids luxury consumption of the elements at the time of application, improves crop uptake efficiency, and increases the formation of reproductive structures. Generally, lower nitrogen and phosphorus levels are used in seed production because these encourage continued vegetative growth. Because most vegetable seed crops are grown in rows, fertilisers are typically applied either in bands in the planting row or as a side-dressing beside the row after establishment.

Weed and Pest Control

The management of weeds and pests in vegetable seed production poses additional problems due to their longer growing season not encountered for vegetables grown for market use. This extended growing period necessitates an increase in residual pesticide activity or additional pesticide applications. In some cases, governmental regulations do not permit the use of the same pesticide for a vegetable seed crop that is used for the market use crop. This is because the approval process is directed at the more economically important value crop.

Some pesticides which control insects also detrimentally affect insect pollinators essential for cross pollination or hybrid seed production and cannot be used in a seed production field. Such restrictions on the use of pesticides in vegetable seed production imply that the safe use of pesticides will be more difficult in the future and greater reliance on crop rotations and biological control will be necessary.

Insects and the timing of their infestations affect vegetable seed production. The Lygus bug prefers legume crops and safflower as its primary host. However, once these crops are harvested, it commonly flies to adjacent carrot seed fields and rapidly creates epidemic infestations. There, it feeds on developing seed embryos and subsequently reduces seed yield and quality. Cabbage looper is a significant pest of brassicas during vegetative development but does not affect seed stalk and pod development. Thus, control of this pest is important during vegetative growth if maximum yield and seed quality are attained. Aphids affect the harvest and subsequent conditioning of brassica seed crops by the development of "honeydew" which adheres to the seed pods and makes separation difficult.

Vegetable seed crops are susceptible to the same diseases as fresh market crops. However, the need to complete their entire life cycle for seed production means that other diseases may also be encountered and require control. Many significant diseases of vegetable seed crops can be controlled by growing the crop in differing geographic locations. For example, the low relative humidity and high temperatures of the southwestern United States reduce the incidence of the seedborne pathogen *Cercospora carotae* in carrot seed fields commonly found in the high relative humidity, low temperature Pacific Northwest. The incidence of downy mildew is reduced by growing long-day onion seeds in the low humidity areas of the desert southwestern United States.

Diseases in vegetable seed fields can also be minimised by separating the seed crop from disease pests either physically or over time. Physical separation occurs by producing the seed crop in an isolated region from fresh market crops or from the disease causing organism. Celery seed fields are isolated from celery stalk production fields to reduce the incidence of late blight caused by *Septoria*. Lettuce seed fields are isolated from leaf and head production fields to reduce the incidence of seed-borne lettuce mosaic virus.

Isolation from the disease causing organism is achieved by maintaining weed-free irrigation canal banks and roadsides which serve as alternate hosts or growing the seed crop where the disease causing organism does not exist. To meet phytosanitary requirements for export, onion seed crops must be grown in regions where certain disease and nematodes are certified to be absent. The lettuce mosaic virus is another serious disease that results in unmarketable lettuce heads. To control this disease, an integrated pest management strategy has been established which focuses on the planting of disease-free seed and control of insects and weeds. Disease-free seed is achieved by planting stock seed in glass houses free of such virus vectors as the green peach aphid (*Myzus persicae*). Commercial seed lots are certified to have not more than one infected seed in 1,000 seeds that have been screened. Other effective measures include control of common weeds that serve as alternate hosts to insect vectors and frequent roguing of lettuce seed fields to remove any plants suspected of carrying the virus.

Some disease causing organisms have relatively short life spans. In these cases, separation of the seed crop from the organism over time reduces disease problems. Alternating cabbage and other cruciferous seed crops in a three year rotation with non-cruciferous crops decreases the incidence of club root caused by *Plasmodiophora brassicae*. Nematodes can similarly be controlled in cucurbit and solanaceous seed crops by three year rotations with crops not from these families. In other cases, diseases can be controlled by storing the seed for a period during which the disease causing organism dies. An example is soft rot (*Erwinia carotovora*) on celery seeds which dies after two years of seed storage. Hot water baths or chemical treatments also control diseases. Hot water treatment is used on a variety of vegetable seeds to control a number of seedborne bacteria and fungi. Many disease causing organisms are carried on the seed coat and most of these are controlled by various topical seed treatments consisting of fungicides, insecticides, and bactericides.

Harvesting, Threshing, and Drying

Vegetable seed crops are divided into two categories based on the type of fruit harvested. Dry-seeded fruits include the brassicas, legumes, and onion and wet-seeded fruits include the cucurbits, melons, and tomatoes. The methods of harvest and extraction of seeds vary dependent on the type of fruit harvested. Of great concern in dry-seeded fruit vegetable seed production is the potential for loss of yield due to seed shattering when harvesting occurs after physiological maturity. To minimise seed loss, two approaches are employed.

Since many dry-seeded vegetables achieve physiological maturity before harvest maturity, they are cut while the plant is still green and placed into rows where the wind further dries the plant and its reproductive structure. This permits more uniform maturation of all dry-seeded fruits and seed quality is enhanced compared to the more costly option of sequential harvest of mature fruits. Later, after the seeds have reached harvest maturity, they are extracted from the seed heads either by hand or mechanically using a combine. The second approach is to visually monitor the development of the dry-seeded fruit and harvest the crop before shattering occurs. For example, onion seeds are very susceptible to shattering once the florets have dried and seed germination and quality is reduced if the scape stalks are cut too early.

A common harvest criterion has been to examine umbels for the first mature seeds. In indeterminate crops such as carrot, seeds mature in sequential order with those on the primary umbel maturing first followed by those on secondary and tertiary umbels. Secondary umbels produce about 50% the quantity of pure live seed of the primary umbel. Thus, harvest of this dry-seeded crop is complicated by the umbel maturation sequence. Seeds produced from the first lettuce flowers also tend to be larger and of higher quality than seeds produced from later flowers.

Seed harvest from wet-seeded fruits is often determined by fruit color. Seed quality in pepper as determined by percentage germination and rate of germination is higher in seeds harvested from red fruits compared to those harvested from green fruits. In tomatoes, maximum seed germination is attained when the fruits are at a mature green stage compared to those which are at the red or fully-ripe stages.

After the fruits are determined to be at the correct stage for maximum seed quality, they are gathered and the seed extracted. In some cases, this is done in one mechanical operation where the fruit is picked from the row, conveyed to a crushing mechanism, and the pulp removed from the seed by washing with water in a rotating drum. For high moisture vegetables such as melons and tomatoes, water is periodically drained from the equipment. For low moisture vegetables such as squash and peppers, water is periodically added to the equipment.

In most cases, vegetable seed companies mechanically harvest crops such as tomatoes, peppers, and cucumber in the field, transport the fruit to a central extraction plant, and remove the seeds from the fruit. In tomato, cucumber, and some melons, a mucilaginous layer around the seed must be removed prior to cleaning. This can be done by two methods. The first, and most common, is the retention of seeds in the crushed fruit pulp for a few hours to more than a day so that fermentation occurs. This process is closely monitored because the fermentation creates heat and mechanical injury to the seeds if continued too long. The second approach is to add sodium carbonate or hydrochloric acid to the crushed fruit extract to degrade the mucilaginous layer. After the fruit is crushed and any mucilaginous layer removed, the seed is separated from the pulp and passed along a water sluice containing gates along its entire length.

Mature high quality seeds sink behind the gates and are separated from the pulp by the running water. After the pulp is removed, the sluice gates are opened and the mature seeds captured on a screen. At this stage, the wet seed must be promptly dried to avoid heating and loss of seed quality with increasing imbibition. To accomplish this, the wet seeds are collected and placed into a spin dryer to remove excess water and the moist seeds are evenly spread on screens and exposed to large volumes of heated air. After drying is completed, the seeds are handled and conditioned in the same way as employed for dry-seeded vegetable crops.

Cleaning and Storage

The physical quality of the seed lot following harvesting,

threshing, and drying must be evaluated to determine whether precleaning is necessary. If the seed lot possesses large amounts of inert material such as stones or plant debris, it should be passed over a scalping machine. If the seed possesses undesired appendages which cause the seeds to clump together such as carrot, then it must be passed through a debearder to ensure free-flow of individual seed units. Impact debearders are also used in multigerm beet seeds to split the fruit into single seed units. After precleaning, vegetable seed crops are conditioned in the same way as other seed crops using the same principles and equipment.

Some additional improvements can be made. For example, greater seed density improves the seed quality of lettuce, tomato, and onion crops. This can be accomplished by placing the seed mass into a solution of specific osmotic concentration. The more dense seed fractions are retained and the more buoyant seeds removed. After cleaning is completed, vegetable seeds are stored using the principles of successful seed storage.

Rather, six of the most important vegetables (onion, tomato, cabbage, lettuce, cucumber, carrot, and sugar beet) are considered. These were selected not only because of their economic value but also because they represent differing botanical families, generally differ in their vegetative/reproductive structure harvest, and illustrate differing seed production practices. Similarities exist in the seed production of related species and many of the principles detailed here can be extended to these crops as well.

Dry Seeds

Onion (Allium cepa)

Onions originated in Iran and Pakistan and were

established staple foods of Egypt and India around 1500 B.C. By the middle ages, they were in Europe and brought to North America by Spanish settlers.

The unique flavor and odor of onions have made them an excellent food source that provides zest to many dishes. Moreover, the numerous ways in which onions are prepared from boiling, frying, stewing, baking, pickling, to eating raw make them a versatile food source. The recent popularity of a health conscious world for salad bars has further increased their agricultural importance. Nutritionally, onions are low in calories (approximately 40 calories for an average size onion) and high in ascorbic acid. They are classified into three agriculturally important groups: common, aggregation, and proliferous. The common onion is the most agriculturally important group. It is a bulb producing type reproduced by seed. The aggregation group is characterised by formation of many lateral bulbs or shoots and is reproduced vegetatively. An example of this is shallots. The proliferous group has bulblets form on the inflorescence and is vegetatively reproduced. An example of this is top-set onions.

Plant development

Vegetative. Onions are a member of the Liliaceae family and are considered a cool-season crop. They are herbaceous biennial plants. The first year of development results in the establishment of a fibrous, shallow root system attached to a base plate of fleshy leaves arising around a small conical terminal stem, a structure botanically termed a bulb. Because the root system is shallow, onions are particularly prone to dehydration injury. The leaf sheaths that project above the ground are sometimes called a false stem to distinguish them from the true stem at the bulb's base. Bulb formation is initiated by changes in day length. Very early types require a 12 hour photoperiod and later types initiate bulbing with a 15 hour photoperiod.

Reproductive. Flowering and seed formation occur in the late Spring or early Summer in response to a vernalisation treatment during the Winter. The onion inflorescence is an umbel that produces 50 to 2,000 flowers. Flowering can be as long as two weeks and is not uniform since the umbel actually consists of aggregations of smaller 5 to 10 flowered inflorescences called cymes. From one to over a dozen flower stalks can be produced on a single plant dependent on its development. More flower stalks are produced on plants grown from bulbs than those grown from seed. The stigma of the flower becomes receptive to pollen after the anthers have released their pollen thereby reducing self-fertilisation. However, open-pollination primarily by insects such as honey bees and leafcutter bees carrying pollen from one flower to the next is common.

Seed

The seed is three sided, 3.0 to 3.5 mm (0.12 to 0.14 inches) long. Its surface is wrinkled, dull and greasy looking, black, and has a depressed scar at one end. Within the seed, the embryo is 6.0 mm (0.24 inches) long and 0.4 mm (0.02 inches) in diameter and is curved in a complete circle. Most of the embryo is comprised of a single cotyledon to which is attached a very small shoot apex, first foliage leaf, and short primary root. It is embedded in a tough thick-walled endosperm. The 1,000 seed weight of onion is 3.6 g (126,000 seeds per pound).

Seed production

Onion seed production can be open-pollinated or hybrid.

Since hybrids have greater uniformity, increased yields, and greater disease resistance, most seed produced today is hybrid. These can be produced either by a bulb-to-seed or seed-to-seed method. The bulb-to-seed method is when onion bulbs are grown, harvested in the fall, stored, and replanted in the Spring. The cold vernalisation requirement is experienced during bulb storage in the Winter. This method is more expensive than the seed-to-seed method but it also permits a grower to easily discard off-types, diseased or otherwise undesirable bulbs. It is practiced primarily for maintenance of onion stock seed.

The seed-to-seed method is when seed is planted in the Summer, carried over through the Winter as bulbs in the field, followed by flowering in the Spring. This method results in higher seed yields than the bulb-to-seed method because there are more plants per hectare (acre) and less time is invested in bulb establishment. While overall costs are less than the bulb-to-seed method, careful attention must be given to pest control. Roguing programs to eliminate off-type plants are also required.

Tillage. An onion field should not be used for seed production more than once every four years. The field must be free of perennial weeds and soilborne diseases. The soil should be friable, fertile, well supplied with humus, and well drained. The soil should be plowed to a depth of 15 to 20 cm (6 to 8 inches), harrowed, and worked until a uniform firm seedbed 10 to 15 cm (4 to 6 inches) deep is established.

Planting. Onions seeds are planted 1 to 3 cm (1/2 to 1 inch) deep in rows 50 to 75 cm (20 to 30 inches) apart dy drills at a rate of 4.5 to 6.7 kg per hectare (4 to 6 pounds per acre). Higher seeding rates and closer row spacings have, in some cases, led to higher yields. Planting should occur as soon as the danger of frost is over. If the fields are irrigated or rainfall is common, row spacings of 30 to 45 cm (12 to 18 inches) can be used.

Fertilisation. The objective of fertilising onion seed crops is to ensure that the onion bulbs reach adequate size before the onset of Winter in order that they can produce vigorous seed stalks the following Spring. Onions respond well to fertilisers because the root structure is limited and shallow. They grow best on soils that are slightly acidic (pH 6.0 to 6.5). Mineral soils should be amended with organic matter to assist in moisture retention and improve the soil physical structure. The equivalent of 56 kg per hectare (50 pounds per acre) of nitrogen should be residually available in the soil or the equivalent applied broadcast before planting. Additional applications of nitrogen per hectare (acre) can be applied in one or two sidedressings to make it easily available to the roots. Both phosphorus and potassium should be broadcast at the rate of 67 kg per hectare (60 pounds per acre) prior to planting. Later applications, if needed, can be sidedressed. Fertilisation of nitrogen and potassium is not recommended at flowering because these make the flower nectar unattractive to bees.

Minor element deficiencies occur in onion, particularly on muck soils. Copper deficiencies result in thin, poorly colored scales that have a poor bulb storage life. This can be corrected with the application of 224 kg per hectare (200 pounds per acre) of copper sulfate. This treatment lasts for several years. Magnesium deficiencies result in leaf chlorosis and are often encountered on muck soils that are alkaline or have recently been limed. Manganese deficiencies cause a gradual interveinal yellowing, twisting, curling, and stunting of leaves. Manganese sulfate at 168 kg per hectare (150 pounds per acre) mixed with the broadcast fertiliser alleviates this problem.

Irrigation. Seed onions produced in the dry West require irrigation because of their shallow roots and high demand for water. An onion crop can use 1 1/2 to 2 1/2 acre feet of water during its development and maturation. Soil moisture should always be above 65% field capacity. In areas where irrigation is practiced, seedbeds are usually irrigated immediately after planting. Then, one to three irrigations are necessary after seedling emergence to ensure rapid, continued growth. As the plants initiate bulbing, irrigation is stopped and the soil allowed to dry.

Weed and Pest Control. Onions are poor competitors against weeds because of their slow growth, small stature, shallow roots, and lack of dense foliage. Weeds interfere with harvest and can become contaminants of the crop seed. The most effective weed control measures include crop rotations, cultivation, and use of selective herbicides. A short fallow period before planting also permits an opportunity to use tillage and non-selective herbicides to control weeds. Cultivation is one of the most successful weed control techniques in onion. Weeds can be destroyed during seedbed preparation. After planting, cultivation every one to two weeks until bulbing occurs is useful. This practice not only suppresses weeds but also loosens the soil after it has hardened following irrigations and rains. Initial cultivations should be shallow since the onion roots are close to the soil surface and damage to the crop can occur.

Harvesting, Drying, and Threshing. Although all seed heads on an onion plant do not mature simultaneously, there still is usually one harvest in the seed field. The seeds are dry enough for threshing when the capsules and small seed stems are brittle and readily break when rolled in the palm of the hand. Threshing is done with a combine.

Conditioning. After threshing, onion seeds are conditioned to eliminate weed seeds, light seeds, and chaff. Initial cleaning is with an air-screen cleaner followed by a gravity table. In some cases, the light separated seed is washed. Heavy, good seeds sink while light, poor seeds float. The washing treatment should not exceed three minutes since the seeds rapidly absorb the water and quality is subsequently reduced. The heavy seeds are retained and immediately spin dried in a centrifuge and further air-dried.

Storage. Onions possess one of the most rapidly deteriorating seeds among major crops. They quickly lose complete viability in less than a year when stored under hot, humid conditions. If properly stored, however, viability of onion seed can be retained for long periods and low seed moisture content and low storage temperature favor longer storage life.

Cabbage (*Brassica oleracea*)

Most wild forms of cabbage are found along the coasts of the Mediterranean Sea which is considered their point of origin. Greek and Roman civilisations were the first to recognise the value of cabbage. Its ease of growth and excellent storability ensured its place as an important crop. Initially, these cultures used cabbage for medicinal purposes. Cabbage juice was employed as a gargle against hoarseness and the leaves covered wounds and ulcers to hasten healing.

There are three major types of cabbage classified according to market demand: fresh market, storage, and sauerkraut. Fresh market types usually form conical to globe heads and have a range of maturities from 50 to 95 days. Storage types are slow in growth and have a firm head with thin, finely veined leaves. Cabbages are used in a variety of ways from raw in salads and soups, boiled or steamed in mixtures of vegetables and cabbage rolls, to pickled in sauerkraut.

Plant development

Because cabbage is a biennial plant, the first year of growth requires the production of a vegetative head followed by flowering and seed production in the second year. It is considered a cool-season crop which grows best between temperatures of 10 and 25C (50 to 77°F).

Vegetative. The cabbage seedling often produces a red hypocotyl with two notched cotyledons and a tap root with lateral roots. The first three leaves have a petiole while later leaves are completely sessile. The leaves are unique from many other crops because they are coated with a layer of wax. The first leaves unfold normally but later leaves only partially unfold and do not expand as much as earlier leaves ultimately forming a head that varies from flat-topped to long-oval in shape.

Reproductive. Cabbage is not photoperiodically sensitive but the apical meristem of plants that have begun to form a loose head must be exposed to four to six weeks of 4 to 7C (39 to 45°F) for vernalisation to occur eliciting production of a reproductive apical meristem. Vernalisation of the seed or seedling does not cause flowering. The flowering process starts from the bottom of the raceme and moves to the top. This often results in racemes on which mature siliques are found on the bottom with unopened flower buds on the top causing an uneven setting and ripening of seed on the plant. The flowers are insect pollinated. When hybrid seed is produced, this normally occurs in a 2:2 ratio of male:female parent rows. After pollination, elongation of a dry pod called a silique containing 10 to 30 seeds is completed after three to four weeks. Dehiscence occurs when the two silique valves split from the point of attachment to the plant upward leaving the exposed seeds attached to the placenta.

Seed

The cabbage seed is almost round, 2.0 to 3.0 mm (0.08 to 0.12 inches) in diameter, reddish brown to grey to black, with a discernable radicle groove and hilum. The seed coat possesses a small reticulated surface. The embryo is large, folded and little endosperm is present in the mature seed. The 1,000 seed weight is 3.6 g .

Seed production

Tillage. Cabbage requires an abundant supply of water. As a result, it typically grows best on soils that are heavy and contain a high level of organic matter. The pH should be between 6.0 to 6.5 for optimum growth. The planting bed for cabbage should be deep and well worked to enable rapid seedling root development and access to water. This can be accomplished by disking the soil. Disk cultivators are useful in preparing seed bed rows about 7 to 10 days before planting because they further pulverise the soil. Additional smoothing of heavy soils, if required, can be done with a harrow.

Planting. Normal cabbage seed production begins with planting in the Summer followed by the half-grown plants over wintering, flowering in the Spring, and maturation and harvest of the seeds in the Summer. All stock seed is tested or hot water treated for black rot prior to planting. Cabbage is planted initially in beds or the greenhouse. The seedlings are then transplanted 45 days later in the field in rows 107 cm (42 inches apart) with a spacing between plants of 40 cm (16 inches) to produce a population of 3,645 to 4,050 plants per hectare (9,000 to 10,000 plants per acre).

Hybrid seed production is more labor intensive. Each parent is normally staked and stringed to allow improved pesticide spraying and facilitate subsequent hand cutting of the inflorescence. The stakes and strings insure that the parents stay separated. The nick between female and male parents must be managed to minimise the number of inbreds and maximise the number of hybrids produced. To accomplish this, bees are placed in the field at 5% bloom. In some cases, plants are topped if one parent comes into bloom earlier than the other.

Fertilisation. Nitrogen during the establishment of cabbage plants leads to a softening of tissues and an increased susceptibility to frost. As a result, nitrogen is usually omitted at planting with phosphorus and potassium applied at rates of 100 and 150 kg per hectare (90 and 134 pounds per acre), respectively. If nitrogen is necessary, it is split as a sidedress in the Fall and added in the Spring just before bolting. This ensures maximum growth of the sidebranches which enhances seed yield. Too much nitrogen in the Spring or Fall enhances soft tissue development, increases the risk of lodging, and promotes extended flowering which may cause increased inbreds in hybrid seed production. Cabbage is not as sensitive to micronutrient deficiencies as are other members of the Brassicaceae.

Weed and Pest Control. Chemical and mechanical weed control programs are practiced with cabbage. Seedbed preparation should eliminate most weeds. When weeds are persistent, however, a preplant or preplant incorporated application of herbicides can be used. The extensive and shallow rooting system of cabbage limits deep cultivation to control weeds, particularly when the soil is hard. Only when weeds are present should large plants be cultivated and all cultivation should be discontinued after heading to avoid further plant damage. The most serious weeds include tiny vetch, wild geranium, and wild turnip which must be controlled or removed prior to harvest.

Irrigation. Cabbage demands a high and consistent quantity of water for optimum development. This is attributed to its large leaf surface area and resultant loss of water through evapotranspiration. The soil water level should be maintained at 60% field capacity or higher throughout crop growth. After bolting, irrigation should be discontinued.

Harvesting, Threshing, and Drying. Generally, each parent is harvested separate from each other. However, in some cases when both parents are stongly self incompatible, they are harvested together. The uneven maturation of siliques on the cabbage raceme and the strong tendency for the seed pods to shatter creates harvesting problems for obtaining maximum yields. For these reasons, harvesting of hybrids is done by hand by cutting the seed stalks with a knife or shears. Open pollinated varieties are swathed by machine.

Conditioning. Removal of remaining chaff and straw from the cabbage seed can be done on an air-screen cleaner. Separating splits from whole seeds is accomplished with a spiral separator. In addition, cabbage seed is normally sized into several sizes for better precision planting.

Drying and Storage. Cabbage seed can remain germinable for four to six years if properly dried and stored at a low moisture content in a low relative humidity (50% or less) environment. The seed should be at a moisture content not exceeding 6% at storage. When drying is necessary, the temperature should not be above 45C (113°F) with lower moisture seeds more resilient to higher temperatures.

Lettuce (Lactuca sativa)

The exact origin of lettuce is unknown. Paintings of a cos-like lettuce are found in Egyptian tombs constructed about 4500 B.C. From Egypt, it is believed that lettuce quickly spread around the Mediterranean basin and was adopted as a popular vegetable by the Greek and Roman civilisations. The Romans further developed differing lettuce types that were broad-leaved, nonheading, nonspiny, had decreased latex content, and were resistant to early seed stalk formation. Eventually, lettuce spread throughout Europe where the present six types of lettuce were selected and produced.

Lettuce is unique among vegetable crops because it is used exclusively as a fresh, unprepared component of salads. There are six morphological types of lettuce: crisphead, butterhead, cos, leaf, stem, and Latin. Crisphead lettuce, often called iceberg, is most common in the United States. This type of lettuce produces large, heavy (up to 1.0 kg/2.2 pound) heads with tightly folded, green outer and white or yellowish inner leaves. Butterhead lettuce forms relatively small, loose heads with broad, oily, crumpled, soft-textured leaves and is commonly consumed in northern Europe.

Cos lettuce, often called romaine, has elongated leaves that are dark or light green with heavy ribs forming a loaf-shaped head and is most popular in southern Europe and the Mediterranean basin. Leaf lettuce produces a variety of leaf types and color but all are characterised by forming a rosette of leaves that do not form a head. This lettuce is used primarily in home gardens because of its ease of production. Stem lettuce is one of the few lettuce types that is not consumed raw. It produces leaves that are coarse, peeled, and cooked. This lettuce is grown primarily in the Orient.

Lettuce is relatively low in essential nutrients compared to other vegetables. It is known for its low calorie level which is attributed to a very high (94 to 95%) water content per leaf. Lettuce contains moderate amounts of vitamins A and C and calcium as well as some phosphorus, iron, sodium, and potassium. Despite its low nutritional value and lack of diversity in food preparation, lettuce is so important as a main component of salads that it ranks fourth in vegetable popularity behind only tomato, citrus, and potato.

Plant development

Like sunflower, lettuce is a member of the Asteraceae family. The cultivated lettuce is closely related to and can easily cross with the common wild lettuce (*Lactuca serriola*). Both produce high levels of latex in the stem and leaves and this trait is responsible for the derivation of the generic name Lactuca meaning "milk-forming". Lettuce is a cool-season crop that requires abundant moisture and ample sunlight for optimum development. Cool nights are important to establish normal head development and high temperatures promote seed stalk formation.

Vegetative. Lettuce is an herbaceous annual that differs from its wild-type relative by producing a rosetted, short-stemmed vegetative plant. The leaves vary widely in shape from round to obovate with varying degrees of cupping and indentation. The root system develops a deep, penetrating taproot with the major absorbing surface being in the top 30 cm (12 inches) of soil.

Reproductive. Lettuce varieties are either long-day sensitive or day-neutral and high temperatures (above 18C/64°F) promote bolting (a change from the vegetative state to flower formation). In most cases, the seed stalk or inflorescence rapidly emerges from the head or leaf structure. In some cases, particularly with crisphead lettuce, the head is so tight and firm that the seed stalk is unable to emerge from the head. To enhance seed stalk emergence, three methods are used: 1) deheading - physical removal of head leaves surrounding the seed stalk, 2) slashing and quartering - cutting the seed head at its apex in an "x-shape" that loosens the tight-fitting leaves over the seed stalk, 3) application of growth regulating chemicals - application of gibberellic acid (20 to 500 ppm) to lettuce leaves at the three to five-leaf stage promotes bolting.

Seed

The lettuce fruit (seed) is an achene. It is lanceolate, flat, and the cream- to back-colored fruit wall has about 8 lengthwise ridges. The achene is 4 to 5 mm (0.16 to 0.20 inches) long, beaked, and topped with a number of large soft, capillary, white or brown, delicate pappi. The 1,000 seed weight of lettuce seed is 0.6 to 1.0 g (453,000 to 755,000 seeds per pound) dependent on cultivar.

Seed production

Approximately 70% of the lettuce seed in the United States is produced in California. This is due to the availability of favorable soils, climate, dependable water supply, and access to available labor since this crop requires a high labor input.

Tillage. Lettuce can grow on soils that range from fine sandy loams to clays to mucks but they must have a high organic matter content and adequate nutrients available in the upper 25 to 30 cm (10 to 12 inches) of soil. Since calcium is an important element for rapid leaf growth, the soil should be at a pH of 6.0 to 6.5 to ensure its availability. Typical vegetable tillage practices should be followed. The soil should be plowed to 30 cm (12 inches) as early in the season as possible. The seedbed is then smoothed by disking or harrowing and then leveled with a spiked harrow. When irrigation is necessary, elevated seed beds are prepared and then dragged flat.

Planting. High quality lettuce seed should be used in planting for greater uniformity of growth. It is often pelleted to facilitate precision planting. The minimum soil temperature required for germination of lettuce seed is 1.7C (35°F) and the optimum temperature is 24C (75°F). High temperatures of 33 to 35C (91 to 95°F) inhibit germination. When lettuce is planted under high temperatures, it is often planted at night and irrigated to induce germination before the high day temperatures are experienced. The seeds also can be osmoconditioned in growth regulators such as kinetin and gibberellic acid and redried, or pregerminated in a gel then planted by fluid drilling.

Fertilisation. On mineral soils, nitrogen is usually limiting and should be applied at planting, after thinning, and at bolting. Phosphorus and potassium should be applied at planting because of the crop's inability to extract these elements from deeper portions of the soil. Muck soils have a lower nitrogen requirement but require phosphorus and potassium.

Weed and Pest Control. Crop rotations with tomatoes, cucurbits, sweet corn, spinach, beets, and carrots assist in minimising the buildup of lettuce weeds and diseases. Weed control is accomplished by both cultivation and herbicide application. This practice cuts the weeds at or immediately below the soil surface, leaves a shallow mulch on the soil, and protects the small developing roots of the lettuce plant from mechanical damage. It may still be necessary for hand weeding between rows at later stages of crop development. Common weeds of lettuce fields are lambsquarter, pigweed, henbit, chicory, groundsel, sowthistle, and chickweed.

Lettuce is prone to a host of pathological disorders. From the perspective of the seed producer, lettuce mosaic virus is the most serious of these because the virus is carried in the seed. Infected seeds germinate to produce plants that are stunted, have yellow leaves, and often fail to produce a harvestable head. More importantly, sap sucking insects such as aphids feed on infected plants and serve as vectors for dissemination of the disease to healthy plants throughout the field. Seed production on infected plants can be reduced by 68% and seed weight by 62%. The virus is transmitted to 1 to 15% of the seeds maturing on the infected plant. As a result, basic seed stocks are often produced in insect-free structures or areas where temperatures are too high for aphid attack.

Most major lettuce producing areas require planting only of mosaic-indexed seed set at some local arbitrary level such as 0.5%. This level is determined by seed companies planting the seeds in flats in greenhouses and growing the seedlings to the three to four leaf stage. Infected seedlings are stunted and lighter than healthy seedlings. The use of mosaic-indexed seed has been very successful. Estimates of reductions in losses of 95 to 100% have been reported as a result of this practice

Irrigation. From seedling to harvest, lettuce demands a high quantity of water. This can be provided either by sprinklers or irrigation. The usual practice is to supply early water by sprinklers and then shift to furrow irrigation when the plant reaches the rosette stage. This allows less use of water since the sprinklers can be metered to permit moisture penetration to 30 cm (12 inch) depth per setting while the lettuce plants are small. Irrigation, however, enhances the level of water related pathogen problems. Irrigated fields should be tiled so that the water can drain through the soil thus minimising waterlogging conditions.

Irrigation for lettuce seed production is generally stopped after flowering. In particular, sprinkler irrigation after flowering causes loss of seed yield since the falling water droplets dislodge the maturing seed off of the flower receptacle.

Harvesting and Threshing. A period of 12 to 21 days is required from flowering to mature seed formation. However, because of the nature of the flowering response, not all seeds mature simultaneously on the plant. As a result, lettuce plants are harvested when approximately half of the seed heads have "feathered" (seed pappus is fully expanded and dry). They can be either machine or more commonly hand harvested.

Conditioning. Harvested lettuce seeds can be as much as one and a half times their dry weight due to contamination with high moisture flower and vegetative plant parts. These must be removed by cleaning as soon as possible to minimise seed deterioration. Lettuce seed is often cleaned on an air-screen cleaner followed by upgrading on a disk separator or an indent cylinder.

Storage. If properly stored, lettuce seeds retain their viability for four to five years. Seeds should be at a moisture content no higher than 7%. Storage deterioration of lettuce seeds is often manifested by an abnormality known as "cotyledonary necrosis". Deteriorated seeds exhibit reddish necrotic spots on the cotyledons. The cause of cotyledonary necrosis is still not known.

Carrot (*Daucus carota*)

Carrots originated in Afghanistan where they expanded to the eastern Mediterranean area. Arabs brought carrots with them during their conquests of Europe in the 10th to 12th centuries. European settlers brought carrots to the New World in the 1600s. Today, carrots are grown worldwide and its wild relative, Queen Anne's lace, is equally well known.

Plant development

Carrots can be grown almost anywhere as long as the growing season is relatively cool. Their optimum growth occurs between 16 to 18C (61 to 64°F). Botanically, the carrot is an herbaceous biennial that is characterised by vegetative production in the first year of growth followed by reproduction growth after vernalisation in the second year.

Seed. The seed is fawn to brown, oblong, 3 to 4 mm (0.12 to 0.16 inches) long and 1 to 2 mm (0.04 to 0.08 inches) wide with one side flattened and the other possessing three to five ribs that are barbed. The embryo is extremely small relative to the size of the seed. The 1,000 seed weight is 0.8 g (567,000 seeds per pound).

Vegetative. Carrots form a vegetative structure consisting of an above ground rosette of leaves and a large, deep tap root. The leaves are double compound with the lower leaves of the rosette being more divided and larger than the upper leaves.

Reproductive. The carrot plant generally requires vernalisation for flowering. This effect is not related to root size since even small rooted plants, if vernalised, will bolt and flower. After vernalisation, the rosette bolts to form a stem that is 61 to 122 cm (2 to 4 feet) high with branches that produce flowering heads or umbels. The appearance of umbels is not uniform. The first and largest umbel to flower is the last umbel on the main flowering stalk and is known as the primary or king umbel. Secondary umbels are formed at the terminus of branches from the main flowering stem and flower in a sequence from the top to the bottom of the inflorescence. Tertiary umbels originate on secondary umbel stems.

Seeds produced on primary and secondary umbels vary in quality. Primary umbels produce seeds that are heavier, more mature, and of higher quality (germination percentage is higher) and vigor (germination rate is faster) than secondary umbels. As a result of this relationship, many seed companies have increased planting density in order that more primary umbels are produced for harvest. In other cases, to enhance seed yields, seed companies have established management practices that focus on harvesting the king, secondary, and tertiary umbels. This allows them to grade seeds into various sizes from large to medium to small.

Hybrid carrot seed production is possible. Male sterile lines have been discovered and in hybrid seed production the common ratio of female to male rows is 6:2 or 6:4. Unfortunately, the production of hybrid seed has required the intensive selective inbreeding of carrots for several generations resulting in inbreds with reduced plant vigor. As a consequence, yields of planted hybrid seed have not been as high as open-pollinated lines. In addition, the male sterile flowers produce smaller petals that are less attractive to pollinating bees causing lower seed set and reductions in seed yield. These factors have led to higher costs for hybrid carrot seed. This, in conjunction with lower crop yield, have continued to make open-pollinated lines the preferred choice of planting seed among carrot growers. However, the commercial fresh market acreage in the United States of the Imperator carrot varieties is almost entirely with hybrid seed. Processing carrots are still dominated by open-pollinated varieties.

Seed production

Carrot seed production used to be located primarily in California. However, some years prevailed when the necessary vernalisation temperatures (below 7C/45°F) to induce flowering were not experienced or high temperatures were encountered during seed set that led to empty seeds. Today, most of the carrot seed production occurs in the Columbia Basin of Washington, Madres area of Oregon as well as Idaho which are ideal desert environments for maximum seed yield and quality.

Carrot seeds are produced by two methods: seed to seed and root to seed. In the seed to seed method, the seed is planted direct in late Summer, vernalised in the Winter, flowers initiated in the following Spring, and seeds harvested in late Summer. In the root to seed method, roots raised in beds usually in an area away from commercial production are removed from the soil in the Fall, examined for trueness to type and stored at 1C (35°F) at 95% or higher relative humidity to reduce storage rots in the Winter, and then replanted in the Spring. The plant bed is used as a backup for direct seeded production in the event of Winter kill. Some types of carrots such as high sugar types do not overwinter well so they may be grown in a plant bed. In both cases, the process of removing plants, checking for off-types, storage, and then replanting is labor intensive and expensive. Even so, the root to seed method is required by most seed companies for new varieties, new inbreds and stock seed to allow selection for root characteristics. The seed to seed method, however, is preferred among seed companies for large commercial seed production.

Tillage. Fields are placed according to isolation distances established for each type of carrot. Carrots grow well on most soils but the best growth occurs on deep, friable soils such as sandy loams. Carrots tolerate a wide range of soil pH. Because seedling establishment is a problem in carrot seed production due to the seed's small size and slow germination rate, particular attention is give to tillage operations prior to planting. The seedbed must be clean of weeds, well pulverised, and compact to ensure ready flow of soil water to the seed and yet minimise puddling or crusting after a heavy rain.

Planting. If planting is done for the seed to seed method, row spacings of 50 to 90 cm (20 to 35 inches) are used and 2 to 3 kg of seed planted per hectare (1.8 to 2.7 pounds per acre). If the seed is mechanically precision planted, it is often coated using a ratio of 3:1 finished weight over the raw seed. Seeds should be planted 1.3 to 1.9 cm (0.5 to 0.7 inches) deep every 7 to 10 cm (3 to 4 inches). If the root to seed method is used, the transplants are placed every 20 to 30 cm (8 to 12 inches) apart in rows that are 75 to 90 cm (30 to 36 inches) wide.

Higher plant densities are desired in carrot seed production. Such a practice results in a concentration of umbels in the upper part of individual flower stalks causing a higher proportion of primary to secondary umbels and a more uniform period of seed maturation from umbel to umbel. Normally, bees are used for open pollinated varieties and are required for hybrid seed production. Usually 2 hives per hectare (5 hives per acre) are required.

Fertilisation. Because of the wide range of soil types in which carrots grow, a soil analysis provides the best recommendation for fertiliser requirements. As a general rule, the relative proportions of 1-2-2 are used. Less nitrogen is needed on muck soils compared to mineral soils. The nitrogen should be applied as ammonia since other forms cause serious burning. The fertilisers are best applied in a band and to the side of the seed at planting. Vegetative growth is further encouraged, particularly under irrigated conditions, by a second side dressing later in the Summer. In the Spring of the second year, another nitrogen application will stimulate vigorous plant growth prior to flowering.

Weed and Pest Control. Weed control in carrots is almost totally dependent on chemicals. This is because any physical alteration of the firm seedbed dislodges the small seed from the soil and results in poor emergence. In the past, herbicidal solvents such as Sotddbard's solvent at 702 to 749 liters per hectare (75 to 80 gallons per acre) were applied after the first true leaves were formed and before the four-leaf stage. This treatment delayed weed growth until the carrot's foliage was large enough to shade competitors. Another approach was to use shielded flame throwers to burn the weeds. Today, weed control in carrots is practiced using a combination of approved herbicides. Weed seeds that are particularly troublesome to remove during conditioning include canada thistle, nightshade, and barnyardgrass.

Irrigation. All carrot seed production is irrigated. Carrots require an abundant and well distributed water supply for optimum growth. The primary approach is to use rill (flood) irrigation. Fields irrigated overhead should be evaluated for the presence of additional fungicidal sprays. The water table should be kept at 76 to 91 cm (30 to 36 inches) below the soil surface. In inorganic soils, the water can be supplied by sprinklers and furrows. Sprinkler irrigation is most effective during the early stages of germination and emergence when the water supply does not have to be extensive. With later vegetative growth, sprinkler irrigation should be discontinued because of the increase in foliar diseases encouraged by the wet leaves.

Harvesting and Threshing. For hybrid seed production, the male rows are removed and turned into the soil prior to swathing. Harvesting of the seed crop is done by machine. This is initiated when the earliest maturing seed on the primary umbel turns brown and the umbel is brittle. Because of the uneven maturity of the seeds on the umbels and the tendency for shattering, umbels can be sprayed with a polyvinyl acetate adhesive to reduce seed loss. This process, however, is not done in large scale seed production. Harvesting early in the day while dew still remains on the umbel is also helpful. The fields are swathed by machine into windrows. After 3 to 5 days of drying, the crop is havested with modern combines at a seed moisture content of 7%.

Drying and Cleaning. The small size of the carrot seed makes it very responsive to equilibration with a low relative humidity environment. As a result, carrot seed is seldom dried unless wide ranges in seed maturity exist. Since the barbs on the seed reduce flowability and increase seed volume, they are often removed by a debearder. Cleaning of the seed is accomplished using an air-screen cleaner and indent cylinders. Carrot seed is sized into varying lengths and widths for improved precision planting.

Storage. Carrot seed should not be stored at moisture contents above 7.0%. At this moisture content and under appropriate temperature and relative humidity conditions, the seeds remain viable for 3 to 5 years.

Sugarbeet (Beta vulgaris L.)

Sugarbeet originated in the Mediterranean. It belongs to the same species as the red or gardent beet. In the middle of the 18th century, it was discovered that sugar from sugarbeet was identical to that found in sugarcane. From that point, countries reliant on sugarcane as their sole source of sugar begainefforts to develop this as an alternative sugar source. France was the first country that to breed a new variety of sugarbeet that could be economically processed for profitable yield.

Plant development

Sugarbeet is a biennial plant that grows vegetatively the first year and produces seed the second year after being vernalised by thermal induction during the Winter. However, when grown as a sugar crop, it is harvested after the first year during which the leaves and the fleshy root are formed. In the first year, it produces a dense canopy of leaves and a large root in which the sugar is stored. In the second year, the sugar is used to form the flowering stalk, flowers, and seed. The leaves are simple with a large blade and petiole. The root is divided into three zones: The top from which the leaves are produced in a dense spiral, the thick fleshy hypocotyl where most of the sugar is found, and the root which becomes a long taproot. The flowers are perfect but incomplete; they have a five part calyx but the corolla is absent. They are produced on a panicle-like spike.

Seed

The sugarbeet "seed" that is planted is actually a single achene, which is a one-seeded fruit with a somewhat inflated, corky outer layer, or pericarp. The traditional natural sugarbeet seed unit was a multi-germ "seedball," or a cluster of two to five achenes formed by the joining of several floral units. When planted, each of the achenes germinated, resulting in three to five plants. The resulting stand of plants was much too thick for optimum root and sugar production and had to be manually thinned at great expense to the grower.

Seed production

It is important that sugarbeet seed producers have a contract with seed companies or sugar companies for seed production. This assures access to varieties that will be in demand by growers. It also assures them of parent seedstocks (i.e., inbred lines) needed to produce seed of different hybrids. Furthermore, it insures seed growers of technical advice needed for managing the planting of parental and restorer lines. Most importantly, it assures them of a market for their harvested seed.

Hybrid seed is produced in two different ways. Most commercial seed is produced by mixing the male-sterile F1 line as the primary female parent with seed of a multi-germ pollinator in a 10:1 proportion. The resulting crop contains a low percentage of seed that is not hybrid. However, since these are large and shaped differently, they are easily separated from the small flattish monogerm seed during conditioning.

The second method is more expensive, but results in a completely hybrid crop. It is used primarily to perpetuate parent stock seed which will be used in three-way or double-cross hybrids. It is produced by growing the multi-germ pollinator in about four separate rows separated by a skip-row from 16 to 20 female male-sterile rows. The male rows are destroyed after pollination and before harvesting the hybrid seed produced on the male-sterile parent.

The male-sterile lines are perpetuated for further use by backcrossing with their type "O" equivalent. This process insures that the progeny will be male sterile. The type O lines are male fertile lines that do not restore pollen production to the offspring when crossed with male-sterile lines. They are comparable to maintainer lines used in producing hybrid corn.

Tillage. Sugarbeet grows best on slightly acidic fertile loamy soils. Sandy soils do not retain adequate moisture between irrigations and heavy clay soils cause problems at harvesting. Fields are plowed to 20 to 30 cm (8 to 12 inches) deep in the late Summer. If sprinklers are not used for irrigation, the fields are leveled for surface irrigation and a firm seed bed prepared by disking, harrowing, and rolling to minimise soil crusting.

Planting. Planting occurs in late August or early September. Seed is planted with a standard drill in rows spaced 51 to 60 cm (20 to 24 inches) apart at a rate of 13.4 to 16.8 kg per hectare (12 to 15 pounds per acre). Seeds are platned from 2.5 to 3.8 cm (1.0 to 1.5 inches deep. The optimum temperature for germination is 15°C (60°F).

The fields must be isolated from other fields of sugarbeets or garden beets that are allowed to produce seed. This assures genetic purity of the resulting seed. For varieties that are similar in characteristics, isolation of 0.8 km (0.5 miles) may be adequate. However, for elite seedstocks or very dissimilar varieties, isolation of 3.2 to 4.8 km (2 to 3 miles) may be desirable. It is important that the guidelines of the seed company for each variety be closely followed.

Fertilisation. Fertility practices vary greatly between seed production areas. In Oregon, the usual practice is to apply 268 kg per hectare (240 pounds per acre) of nitrogen and 112 kg per hectare (100 pounds per acre) of phosphate, along with 56 kg per hectare (50 pounds per acre) of sulfur and 5.6 kg per hectare (5 pounds per acre) of boron. Another 179 kg per hectare (160 pounds per acre) of nitrogen are side-dressed in the Spring, often in two applications.

Weed and Pest Control. Good weed control includes cultivation and chemical weed practices in addition to hand weeding. Sugarbeet grows rapidly and soon shades most weeds and prevents them from establishment. Cultivation should be used wherever possible. However, chemical control is the most common method and many chemicals are registered for all seed production areas. Finally, hand weeding is an important additional weed control practice. The most common weeds of sugarbeet seed fields are pigweed, lambsquarter, barnyardgrass, mustard, and wild oats.

Rarely can sugarbeet seed crops be grown continuously on one field. Rather, they are often rotated with cereals, beans, vegetables such as tomatoes and lettuce, and forage legumes to control diseases and root-knot nematodes. Black root is a fungal disease that kills germinating seedlings and reduces plant vigor so that uneven stands and poor yields result. Seed treatment is effective against this disease. The virus disease, curly top, is a particular problem for sugarbeet seed producers in the intermountain area. This disease is controlled almost exclusively through the use of disease resistant varieties.

Harvesting. Sugarbeet seed shatters readily as it approaches maturity. Shattering is minimised by windrowing the crop before the seed reaches full maturity and allowing it to cure a few days before combining. Windrowers are equipped with vertical cutter bars on each side to enable the crop to be cut with a minimum of shattering. In drier areas such as Arizona and Utah, this requires about 7 to 10 days compared to about 10 to 14 days in the Willamette Valley. The cutter-bar should be raised high enough so the plants are gently laid on the ground so air can circulate through the windrow and allow aeration and drying. Some of the mature seed will still shatter during windrowing, however most will fall onto the windrow and can be recovered during combining.

Conditioning. The harvested seed is first conditioned over an air-screen cleanter to remove inert material as well as multi-germ seeds which contaminate the lot. This is followed by a draper belt which further removes stems and remaining inert matter. The remaining multi-germ seed is removed by width or thickness graders which allow the small monogerm seed to fall through, but reject the large multi-germ seed. The monogerm seed are further conditioned by passing between two horizontal disks, one of which rotates. This action rubs the corky tissue from the seeds and reduces them to small, uniform size and shapes. Further screening can remove all chaff and inert material. Finally, the seeds are conditioned over a gravity table which removes light weight, low quality seed from the seedlot. The seed is then treated with a fungicide and an insecticide.

Wet Seeds

Tomato (Lycopersicon esculentum)

Tomatoes originated in Peru and Bolivia as small, wild fruits. Migration of local natives brought the tomato to Central America and Mexico where large, domesticated fruits were developed. Archaeological evidence reveals that tomatoes were a popular fruit of the Aztec and Toltec Indians who planted them with their corn. Spanish settlers to the New World carried tomatoes back to

Europe where they quickly became established throughout the continent. Initially, tomatoes were grown as ornamentals because of their colorful, bright red fruits and the assumption that the fruits were poisonous.

Plant development

The tomato is a warm-season herbaceous perennial that requires three to four months from seeding to production of fruit. It grows best at temperatures between 18 and 29°C (65 and 85F), is often killed at temperatures below freezing, and does not grow at temperatures above 35°C (95F).

Seed. Tomato seeds are 3 to 5 mm (0.12 to 0.20 inches) in size, flat, reniform, and have a grey, slightly hairy seed coat. The embryo is large and coiled in a limited, fleshy endosperm.

Vegetative. Tomato plants are typically viny, prostrate, and are either determinate or indeterminate based on whether the apical stem terminates in an inflorescence. Most shoots form in the axils of leaves. The tomato plant produces a deep tap root with extensive secondary roots.

Reproductive. Tomato flowers are yellow and formed on racemes. The stamens are fused with the lower part of the corolla forming a staminal cone. Generally, pollen release occurs prior to elongation of the style through the staminal cone resulting in self-pollination. However, temperature has a major effect on this process. If temperatures are below 15°C (60F) or above 29°C (85F), pollen release is restricted resulting in incomplete fertilisation of ovules. This causes collapsed fruit walls and formation of deep indentations in the fruit, a phenomenon called "catface." Hot, drying winds also cause tomato flowers to abscise from the plant. The best shaped and fullest tomato fruits result from optimum fertilisation of ovules. The mature tomato fruit is botanically a berry. Fruit color varies from the common red to yellow, pink, and orange.

seed production

Seed production in greenhouses has occurred on a limited scale. As a result, most of the following discussion will consider tomato seed production as practiced under field conditions.

Tillage. Crop rotations are useful in tomato seed production because of the crop's susceptibility to a variety of diseases. Rotations with soybeans, sugar beets, wheat, and beans have been successful. One of the most common rotations in the California San Joaquin Valley is with cotton. Fields used in tomato seed production should be Fall plowed to a depth of 20 cm (8 inches). This practice not only improves the physical structure of the soil but also encourages the decay of roots and other plant parts thereby enhancing the level of soil organic matter. If organic matter remains low, manure can be applied at a rate of 13,440 to 17,920 kg per hectare (6 to 8 tons per acre) and incorporated into the soil at the time of Fall plowing.

Planting. Tomatoes can be planted either by direct seeding or transplants. When 152 cm (5 foot) rows are used, seed is planted to give a spacing of 20 to 30 cm (8 to 12 inches) or 30 to 40 cm (12 to 16 inches) if double rows are used. Planting should not begin until the soil temperature has reached 10C (50°F) since germination will not occur below this temperature. The optimum range of soil temperatures for seed germination is between 15C and 29C (60°F to 85°F).

Direct seeding is accomplished by precision planters that equally space the desired number of seeds in the row. Seeds should be planted at a depth of 1.3 to 2.0 cm (1/2 to 3/4 inch) to obtain a population of 24,700 to 49,500 plants per hectare (10,000 to 20,000 plants per acre) in single rows. Planters should be modified to allow the application of anticrustants (keep the soil from crusting and thereby facilitating uniform seedling emergence) and starter fertilisers applied either in a solid or liquid form.

Transplanting is another important planting approach used in tomato seed production. Transplants can be obtained either from local greenhouses or can be from transplants produced using grower seed. They are at the correct stage for planting when the first pair of seedling leaves appear. When possible, transplanting should be done either during the afternoon or on a still, cloudy day to avoid transplant shock. The transplants should be thoroughly watered before planting and a starter fertiliser high in phosphorus added. Transplants every 40 to 45 cm (16 to 18 inches) in rows 168 cm (66 inches) apart provide a population of 14,800 to 19,700 plants per hectare (6,000 to 8,000 plants per acre).

Fertilisation. While tomatoes grow on almost any soil, soils that are sandy, light, well-drained, and high in organic matter and mineral nutrients produce the best quality and highest fruit yields. Optimum soil pH should range from 6.0 to 6.5. Tomatoes require an abundance of the three major elements: nitrogen, phosphorus, and potassium. Adequate nitrogen is important to enhance foliage growth which has a major bearing on crop maturity and protects fruits from sunscald. Phosphorus influences fruit quality by stimulating vigorous root growth that enables more nutrients to enter the plant thereby promoting sturdy stem growth and healthy leaf formation. Tomatoes use large amounts of potassium. This element is important in stimulating early plant growth and regulating normal carbohydrate and protein metabolism.

Fertilisers are generally applied broadcast before Fall plowing. All the phosphorus and potassium and one-half the nitrogen are added at this time. Between the time the plants have two to three leaves and when the flowers are first observed, a side-dress nitrogen application is useful. Nitrogen should not be applied after fruit formation because it stimulates vegetative growth.

Weed and Pest Control. Mechanical and chemical control of weeds is practiced in tomatoes. Cultivation to eradicate weeds begins soon after seedling emergence. Early cultivations are close to the seedling and 2.5 to 5.0 cm (1 to 2 inches) deep. Later cultivations are further from the plants and shallower to minimise injury to rapidly spreading surface roots. Chemical control also involves application of an array of pesticides specific to the weed problem. Weeds that are difficult to control in tomato fields include nightshade, pigweed, lambsquarters, green foxtail, wild oats, nutsedge, smartweed, and morningglory.

Harvesting. Tomato fruits can be hand or machine harvested. In the case of hybrid seed production, the marked or tagged fruits are always hand harvested. All mechanical tomato harvesters are based on a "once-over" principle in which the entire tomato plant is harvested. They have four components: pick-up mechanism, fruit and vine separating area, fruit sorting area, and container loading mechanism.

Cucumber (Cucumis sativus)

Cucumber likely originated in India because a close weedy relative, *Cucumis hardwickii,* has been located there. Evidence exists that the cucumber was cultivated in western Asia over 3,000 years ago. From there, it spread westward into Egypt and was introduced throughout Europe by the Greek and Roman empires. Columbus and Spanish settlers brought cucumbers to North America where they quickly became established as a popular crop.

Plant development

Cucumbers are considered a warm-season crop which grows best at average daily temperatures from 18 to 24°C (65 to 75F). Young plant development can be severely retarded if exposed to frost. Cucumbers are noted for requiring the shortest interval from planting to fruit production of any major vegetable crop.

Seed. The seed is flat, oval-ovate in shape, 10 to 15 mm (0.4 to 0.6 inches) long, pointed at the hilum end and rounded at the opposite end. The seed coat is smooth, white to black, with a marginal groove on each side near the base. Internally, the embryo is surrounded by a thin perisperm. The embryo is large, having completely digested the endosperm, and is composed of two leafy cotyledons, each with a prominent midrib, attached to a small, straight embryonic axis. The 1,000 seed weight is 25 g (18,000 seeds per pound).

Vegetative. The cucumber is an herbaceous, prostrate, annual vine that produces from three to eight angled stems covered with short, stiff hairs. The leaves are simple, large, triangular and palmately five-lobed with the middle lobe being sharply pointed. The root system produces a taproot that penetrates more than a meter (39 inches) into the soil with numerous shallow, lateral secondary roots. The lateral root extension is often equal to that of the above ground shoot growth.

Reproductive. Cucumbers are typically monoecious plants that produce imperfect yellow male and female flowers. The flowers are yellow and 2 to 3 cm (0.8 to 1.2 inches) in diameter. Male flowers generally appear first and in clusters in the leaf axils. Female flowers occur after the male flowers appear with usually one flower found in each leaf axil. The appearance of either male or female flowers can be altered by the environment. Male flowers are generally formed under long days while female flowers occur under short days. The flowers are cross-pollinated by insects. Bee hives are often placed around the periphery of cucumber seed fields to enhance pollination. One hive may pollinate up to 1.2 hectares (3 acres) of cucumbers.

Seed production

Tillage. Cucumber seed fields should be rotated from other cucubits such as melons. Cucumbers can be grown on almost any soil. Cucumber fields should be plowed deeply and early so that the field is settled prior to planting. Additional disking and harrowing operations may be necessary to smooth the seedbed. If irrigation is used or if the soil is poorly drained and heavy, raised beds are necessary. This is accomplished using a lister (a double moldboard plow) that makes the furrows 20 to 25 cm (8 to 10 inches) deep and the beds are formed by back-furrowing with a turn plow. A raised seed bed offers two important benefits: 1) the vines and fruits remain above the water level and 2) the raised beds are warmer; providing additional protection against frost injury.

Planting. The time of planting is delayed until the danger of frost is over. Because of the irregular, flat seed shape, most cucumber seed is coated so that precision planting can be accomplished. Direct seeding using drills into rows that are 91 to 122 cm (3 to 4 feet) apart at a rate of nine seeds per meter (three seeds per foot) is practiced.

After emergence, the seedlings are thinned by cutting (pulling creates injury to the developing roots of the plant to be retained) to one plant per foot. When the initial vegetative leader stem has produced three to five leaves, it is removed by pinching to encourage growth of the remaining two lateral stems and their fruits. This results in the development of four or five fruits per parent plant.

Fertilisation. Optimum growth of cucumbers requires soils with high organic matter and a pH of 6.5 or above. The organic matter is often supplied by green manure crops that are turned under before they reach maturity and become woody. It is difficult to provide a specific fertilisation recommendation because of the diversity of soils on which cucumbers successfully grow. However, a fertilisation regime of 1-2-2 is suitable before planting under non-irrigated conditions. Under irrigated conditions, the ratio should be 2-1-1 with half of the nitrogen applied a month after emergence to to replace that lost from leaching.

Weed and Pest Control. Chemical weed control in cucumber is expensive and uncertain so the most effective practices have concentrated on land management and mechanical cultivation. One approach has been to use a stale seedbed in which the field is left untouched until two weeks prior to planting. At that time, the site is sprayed with a contact herbicide to kill all plants followed by direct seeding of cucumber. Another approach has been to plow the killed plants under the soil following herbicide application to increase organic matter. Cultivation 5 to 10 cm (2 to 4 inches) deep begins as soon as seedlings have emerged. Later cultivations, however, must be shallow to avoid damage to subsurface roots and not close to the plant in order to minimise damage to the vine. After vines have covered the ground, further cultivation is not possible.

Irrigation. Cucumbers require high levels of water during vigorous vegetative and reproductive growth. Even in areas where rainfall is plentiful, periodic droughts substantially reduce yields and irrigation is necessary in these instances. While furrow irrigation is preferred, when overhead irrigation is used, it should be applied early in the day to permit the vegetation to dry out prior to nightfall and thereby minimise fruit rooting and foliar diseases.

Harvesting and Extraction. Fruit maturity affects seed quality in cucumber. For maximum seed quality, cucumber fruits must reach full maturity. This is signaled by a change in color from green to yellow and by a withering of the fruit stalk. Fruits can also be examined to determine seed maturity.

Seeds that separate easily from the flesh of a sliced cucumber fruit are ready for harvest. At this stage, the fruits are typically harvested by hand and carried to a central seed extraction, drying, and storage facility for stock seed or small seed lots. For commercial seed, the plants are windrowed and then machine harvested. The slurry is augered into steel bins and then taken to the plant for seed extraction.

Drying. After extraction, the seeds are immediately dried. This is done either in the sun with frequent stirring and turning of the seed or with artificial batch driers that have a rotary paddle to expose the seed to heated air. The initial drying temperature should not exceed 40C (104°F) and the seed should be dried until it reaches a moisture content of 6% prior to cleaning and storage.

Cleaning. Cucumber seeds may still contain slight amounts of fruit debris following drying. This can be removed by an air-screen machine. Light and immature seeds can be separated from mature seeds on a gravity table.

Storage. Cucumber seeds stored at 6.5% moisture content under proper temperature and relative humidity conditions can remain viable for 5 to 7 years.

12

Seed in Present-day Forestry

Forest plantations are a powerful tool in the continuing efforts of foresters to increase productivity per unit area—the only means of reconciling the increasing demands for forest products and services on the one hand with a decreasing area of land available for forestry on the other. A combination of intensive site preparation with the use of uniform, well-grown nursery stock, planted at uniform spacing, increases growth and yield, reduces rotation length, facilitates tending and harvesting operations and improves the quality and uniformity of wood, as compared with natural forest.

Plantations also offer the means of using on a large scale the genetically improved material developed by tree breeders. Although there is no case for the indiscriminate replacement of all natural forest by plantations, their judicious use, by providing an alternative source of forest products, can itself reduce pressure on the remaining natural forest and so help to conserve it as a habitat and a source of genetic diversity.

Not only do plantations have a major role as producers of timber, pulpwood and wood-based panels for forest industries, but fuelwood and pole plantations and farm woodlots are locally important in many countries. Tree-planting is not confined to block plantations. Shelterbelts and dispersed planting for soil stabilisation, habitat improvement, urban and rural amenity or as part of an agrisilvicultural system all benefit the human environment.

With such a variety in planting purposes, it is not surprising that the scale of tree-planting and the variety of species planted continue to grow in so many countries. The current greatly increased interest in agroforestry opens up a whole new range of species for trial. Ability to grow in symbiotic relationship with agricultural crops will be the essential characteristic and will involve criteria such as rooting habit, ability to fix nitrogen and multipurpose uses (food, wood, shelter). Low stature may be beneficial and shrubs may become as important as trees. These new developments will introduce new opportunities and new problems in seed collection and handling.

With a few exceptions, notably among the poplars and willows and in some tropical species of Casuarina, trees are propagated from seed, and the suitability and quality of the seeds have a big effect on the success of the plantations raised from them. The use of sound seed from stands of high inherent quality is widely recognised as the best means of ensuring fast-grown and healthy plantations capable of yielding high quality wood. Seed quality comprises both genetic and physiological quality.

It must be stressed that "good seed" implies seed which is both of high viability and vigour and is genetically well suited to the site and to the purpose for which it is planted. Physiologically good seed may lead to successful establishment of a plantation but this is of little value if it is slow-growing, ill adapted to the site or produces the wrong kind of wood because the provenance or genotype was incorrectly chosen. On the other hand, there is little point in producing genetically improved seed at an increased cost if it is killed by poor handling techniques and has to be replaced or supplemented by inferior seed in order to achieve planting targets. Good seed handling is an essential complement to genetic improvement.

Quantity, as well as quality, of seed production is important. In natural stands variation in the quantity of seed produced affects the forester's decision as to the years in which to collect seed and the trees from which to collect it. More intensive management affords him the opportunity of stimulating heavy seed production in genetically superior crops through deliberate treatment, e.g. thinning in seed stands or a combination of initial spacing, irrigation, fertilisation and thinning in seed orchards. The programmes of seed supply should include measures to improve production quantitatively, as well as improving the genetic quality of the seed and the efficiency of collection and handling.

There is a considerable body of published information on seed handling in the temperate zones but published accounts of tropical experience with forest seeds are scattered and incomplete. While recapitulating the principles of seed handling derived from temperate experience, this guide seeks to illustrate these, as far as possible, by examples taken from tropical species.

At the same time it must be emphasized that there is great variation in seed biology and that certain techniques which are commonly practised with good results in the temperate zones, e.g. stratification or prechilling, may have no useful function whatever in the tropics. It is dangerous to extrapolate temperate experience to tropical species, experience in the dry tropics to rain forest species, or experience in tropical agriculture to tropical forest tree seeds, without testing it in each case.

Both equipment and techniques must be adapted to local species and conditions. For example hard-coated seeds of high natural longevity which are produced regularly every year require no more for storage than a well-ventilated room with protection against pests and diseases. A domestic refrigerator or "deep freeze" may be perfectly adequate for long-term storage if only small quantities are to be used each year. Large units and mechanised equipment may be essential in some countries, but heavy expenditure on such items should be incurred only after a full evaluation of the alternatives.

The chain of seed handling operations is only as strong as its weakest link. Therefore great care must be taken at all stages if seed viability is to be maintained all the way from the parent tree to the nursery bed. If a seed has lost viability in an early stage of the process, the best storage or pretreatment can never bring it back to life. Perfect extraction and cleaning of seeds are a waste of money if they are killed later by incorrect storage conditions or careless handling in transit.

The greatest risks to seeds occur during temporary storage immediately after collection, during transit to the processing depot, and again during transit from the seed store to the nursery. These are the periods when sophisticated equipment is not available and everything depends on common sense and meticulous care to ensure good ventilation and avoid extremes of temperature.

"International exchange of tree seeds is an opportunity to share the world's forest wealth". The widespread success of species such as Pseudotsuga menziesii, Pinus radiata, Eucalyptus globulus and Tectona grandis is a convincing demonstration of the advantages which many countries can derive from the introduction of exotics. International cooperation in seed procurement is now practised by many countries and has long been supported by international organisations such as FAO and IUFRO. It should be increased.

At the same time the movement of seed between countries does involve some additional problems in maintaining both the viability and the identity of the seeds. Phytosanitary regulations, including inspection and treatment, must be complied with in order to reduce to a minimum the risk of importing new and dangerous diseases. But every effort must be made to eliminate all unnecessary delays in transit, including those imposed by customs, plant hygiene and air lines, and to avoid excessive or duplicated fumigation treatments which kill the seeds. Both sender and recipient must be aware of all the current regulations in force in the exporting and importing country and must plan the seed shipment well in advance to ensure smooth and unimpeded movement of the consignment.

Apart from a few well-known species such as Tectona grandis, research on tropical forest seeds has been inadequate in comparison with both the severity of the problems and the large number of species of potential value for plantations. Much has still to be learned. A first step to this is a good understanding of the natural reproductive biology of each species. For species of the dry tropics which survive naturally by means of seedcoat dormancy, storage should present little problem and a modest programme of research should indicate the most appropriate pretreatment to overcome dormancy and induce uniform nursery germination.

The problem of preserving viability in recalcitrant seeds of tropical rain forest species, especially those which will not survive temperatures below about 10° C, is a much more intractable one. A variety of possibilities has been suggested but so far little progress has been made towards a practical method of storage which would be applicable in large-scale afforestation projects.

Seed and Fruit Development

Some knowledge of the biology of seeds is essential to their proper handling. The use of seed for artificial regeneration makes possible a considerable degree of control over the conditions in which it is collected, processed, stored and treated, but the seed's inherent characteristics have been evolved as a result of millenia of adaptation to natural regeneration under local conditions. Knowledge of flowering phenology enables the collector to select the timing and methods of seed harvesting most appropriate to the species, while handling, storage and pretreatment of seed will benefit from a knowledge of how seeds develop in nature.

Pollination and Fertilisation

A seed is a reproductive unit which develops from an ovule, usually after fertilisation. Ovules are borne by both the angiosperms (true flowering plants) and the gymnosperms (which include the conifers). In the angiosperms the ovules are totally enclosed within the ovary, while in the gymnosperms the ovules are "naked", typically borne in pairs on the upper surface and near the base of each scale in a female cone. Since the cone scales remain tightly closed except at the time of pollination and later at seed shed, the term "naked" is a relative one.

Seed development is initiated by fertilisation, the union of a haploid male nucleus from the pollen grain with a haploid female nucleus within the ovule to form a new diploid organism. Fertilisation must be preceded by pollination, the arrival of a pollen grain on the stigma of the female flower in angiosperms or close to the micropyle of the gymnosperm ovule. It is important to distinguish the two separate processes of pollination and fertilisation. In most angiosperms the elongation of the pollen tube is rapid and the interval between pollination and fertilisation is only a few days or even hours. In a few angiosperms (e.g. Liquidambar, some species of Quercus) and many gymnosperms (e.g. Pseudotsuga, Larix, Picea) the interval is several weeks or months, while in other species of Quercus and in many Pinus it is a year to 14 months.

Angiosperm Seed Development

At the time of fertilisation a typical angiosperm ovule consists of one or two protective coats—the integuments—and a central tissue—the nucellus. Often the integuments and the nucellus are clearly differentiated only in the region of the micropyle—the minute pore in the integuments through which, in many species, the pollen tube enters the nucellus. The ovule is attached to the wall of the ovary by a stalk—the funicle.

Meiosis of a mother cell within the nucellus, followed by several mitotic cell divisions, leads to the formation of the embryo sac, a haploid eight-nucleate, seven-celled structure which occupies the central space within the nucellus. When the pollen tube reaches the embryo sac it releases two male gametes. One male gamete unites with one of the nuclei in the embryo sac -the egg cell—to form a zygote which later develops into the diploid embryo plant. The second male gamete unites with two other female nuclei—the polar nuclei—to form a triploid cell which later develops into the endosperm, a tissue which acts as a food reserve for the growing embryo. The remaining five nuclei of the embryo sac (2 synergids and 3 antipodal cells) play no further role in seed development. Successful fertilisation of the egg cell and successful triple fusion with the polar nuclei are both necessary for development of a viable seed.

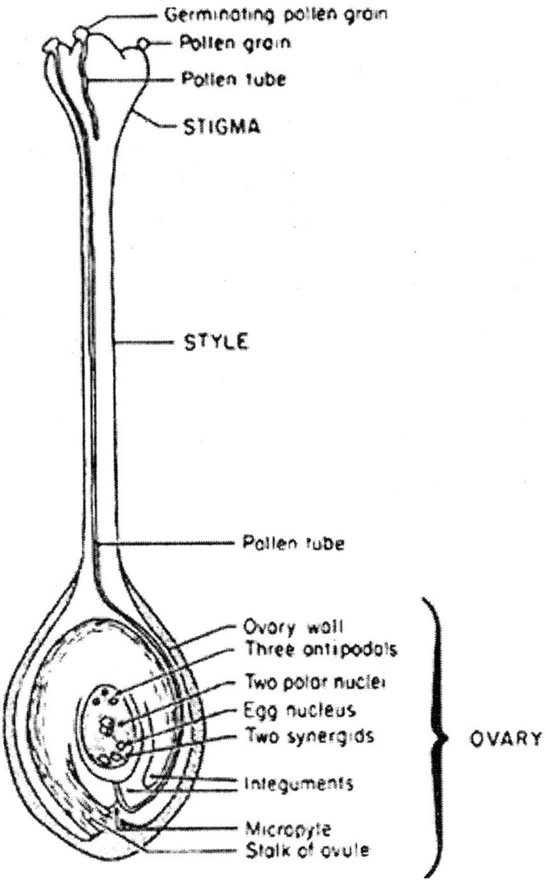

Figure 1. Longitudinal section through a typical pistil just before fertilisation

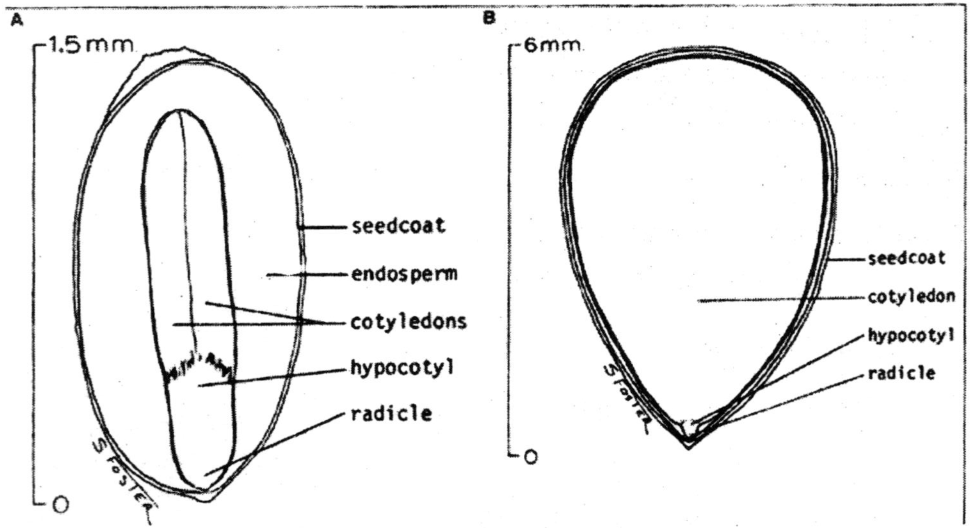

Figure 2. Longitudinal sections through ripe seeds of:- (A) Paulownia tomentosa *showing conspicuous endosperm. (B) Tectona grandis. Endosperm has disappeared and cotyledons occupy almost the entire seed cavity.*

Development of the fertilised ovule into the mature seed involves several different parts. From the outside inwards these are as follows:

1. The integuments of the ovule become the seedcoat of the mature seed. This sometimes consists of two distinct coverings, a typically firm outer seedcoat, the testa, and a generally thin, membranous inner coat, the tegmen. The testa protects the seed contents from drying out, mechanical injury, or attacks by fungi, bacteria and insects, until it is split at germination. But there is great variation in seedcoat among angiosperms.

Examples of different types of fruits:

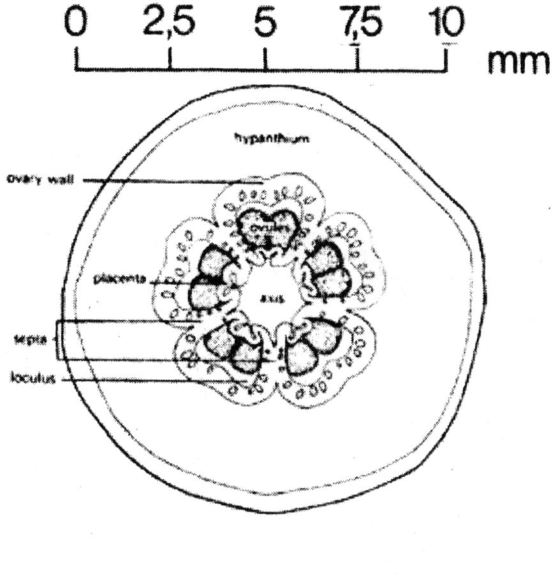

(A) Cross-section of capsule of Eucalyptus preissiana showing loculi, axis, placentae and ovules.

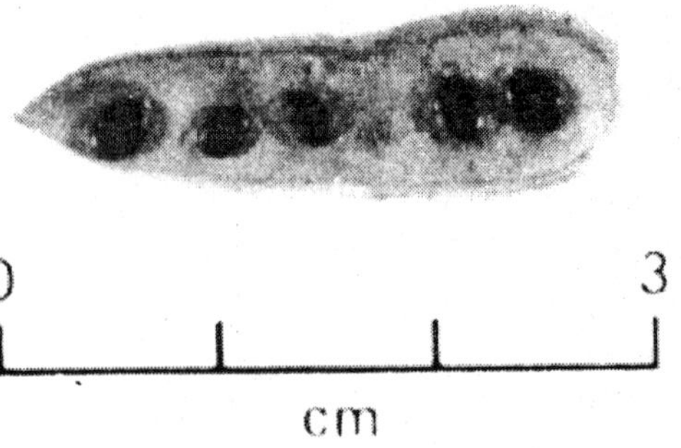

(B) Open pod with seeds of Acacia aneura.

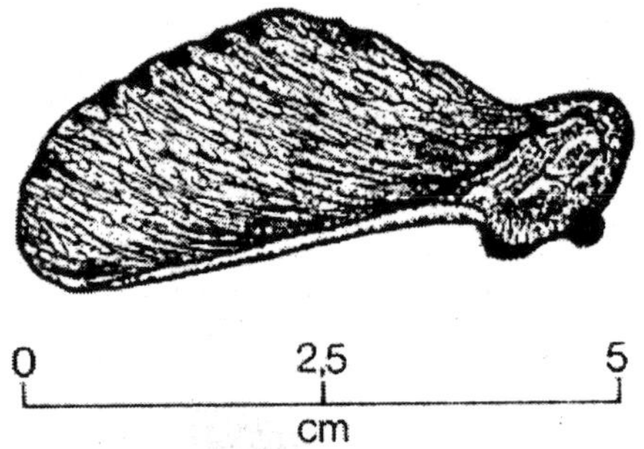

(C) Samara of Triplochiton soleroxylon.

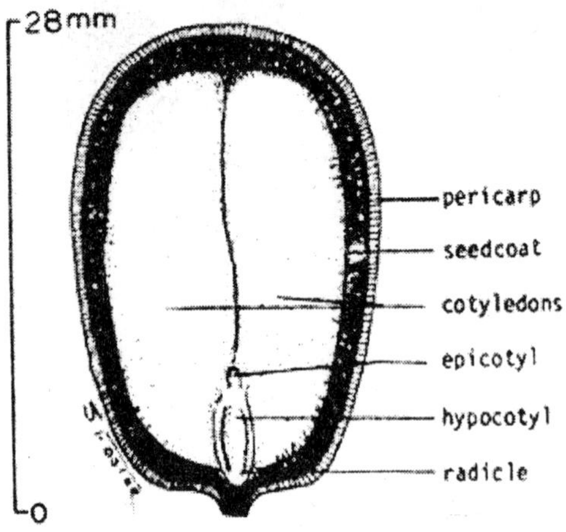

(D) Nut (acorn) of Quercus rubra.

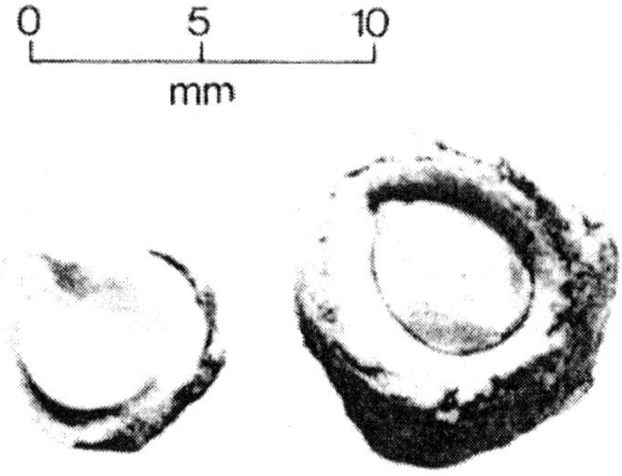

(E) Drupe of Tectona grandis.

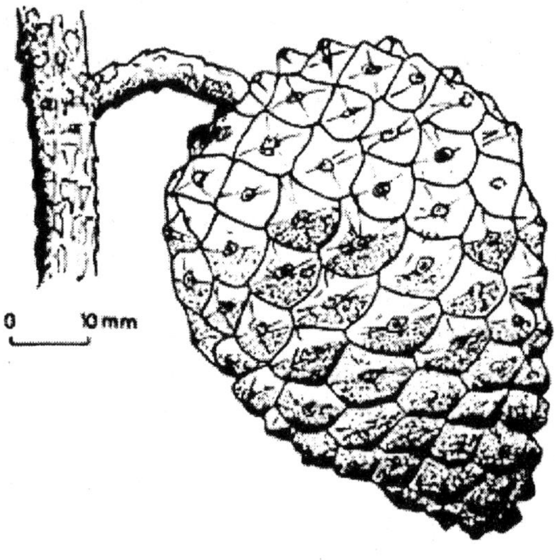

(F) Cone of Pinus oocarpa.

2. The nucellus may persist in some genera as a thin layer—the perisperm -lying inside the seedcoat and supplying food reserves to the embryo. In most angiosperms, however, it soon disappears and its function is taken over by the endosperm.

3. The endosperm commonly grows more rapidly than the embryo during the period immediately after fertilisation. It accumulates reserves of food and its fullest development is rich in carbohydrates, fats, proteins and growth hormones. In some species the endosperm remains conspicuous and still fills a greater part of the seed than the embryo even when the seed is ripe. In others, such as Tectona, the embryo absorbs food reserves from the endosperm during its later stages of development, until the endosperm disappears by the time the seed is mature.
4. The embryo occupies the central part of the seed. Its degree of development at the time the seed is ripe varies greatly according to species. In some it is possible to distinguish all parts of the rudimentary plant—the radicle which at germination will give rise to the primary root, the seed leaves or cotyledons, the plumule from which will develop the primary shoot, and the hypocotyl which connects the cotyledons with the radicle. If the embryo absorbs all the food reserves from the endosperm, the thick fleshy cotyledons commonly become the main organs for food storage and occupy almost the whole of the seed cavity.

Although the storage function within the embryo is normally performed by the cotyledons, in Anisophyllea, Barringtonia and Garcinia it is completely taken over by the swollen hypocotyl which fills the seed cavity; the cotyledons are vestigial or absent. This is also the case in Lecythis and Bertholletia, hence the edible content of a Brazil nut (Bertholletia excelsa) is neither endosperm nor cotyledon but hypocotyl.

In some species the embryo is still small and undeveloped when the seed is ready for dispersal, and an additional period under suitable environmental conditions is needed for maturation of the embryo after seed shed, before the seed can become capable of germination, e.g. Fraxinus excelsior. At its most complex the ripe seed may thus consist of diploid tissue from the mother tree (the seedcoat, including testa and tegmen, and the perisperm), triploid tissue in the endosperm, and diploid tissue of the new genetic combination in the embryo offspring. But both perisperm (nearly always) and endosperm (not infrequently) may be missing.

The essential constituents of all seeds are the embryo, the protective covering of the seedcoat and a reserve of food substances which may be stored in the cotyledons, hypocotyl, endosperm or perisperm. Occasionally more than one embryo may develop in a single seed and such polyembryony has been reported from several genera. It is, however, the exception.

Angiosperm Fruit Development

Development of the fertilised seed is normally accompanied by development of the fruit. In the simplest case the ovary wall becomes thickened to form the pericarp. This may be:

a) Dehiscent, splitting open when ripe to release the enclosed seeds; examples are the capsule (e.g. Eucalyptus), a multilocular fruit derived from a syncarpous ovary, and the leguminous pod (e.g. Cassia), which is derived from a single carpel and spilts along two sutures. The pericarp may be dry, semi-fleshy or fleshy at the time of dehiscence. Semi-fleshy to fleshy capsules are common in the humid tropics (e.g. Baccaurea, Durio, Dysoxylum,

without fertilisation of an egg. Such fruits, called parthenocarpic fruits, occur in several genera of forest trees including Acer, Ulmus, Fraxinus, Betula, Diospyros and Liriodendron. Mature fruits do not invariably indicate mature seed, still less can the number of sound seeds be predicted from the number of fruits. In Tectona the number of sound seeds per fruit can vary between 0 and 4 and still greater variation is possible in other genera.

Seed Dispersal in Angiosperms

There is thus an immense variety among angiosperm fruits. Much of it is related to the need for seed dispersal. Survival and growth of young seedlings under the parent tree is often difficult, because of lack of light and intense root competition. Dispersal over a wide area can ensure that some seeds find conditions suitable for germination and survival, even though the vast majority will perish from the effects of harsh site conditions, competition or destruction by animals or disease.

Dispersal by wind is assisted when the seeds are very light and small e.g. Eucalyptus, or when either the seedcoat (Salix, Ceiba, Dyera) or the pericarp (Triplochiton, Pterocarpus, Koompassia, Casuarina, Fraxinus) possesses wings or hairs which serve to prolong flight. Fruits may also be winged by the enlargement of persistent sepals (most Dipterocarps) or persistent petals (e.g. Gluta, Swintonia).

The distance of seed or fruit dispersal by wind depends not only on the weight and type of dispersal unit but also on the local wind conditions and the exposure and isolation of the mother trees. Studies of the winged fruits of Shorea contorta in the Philippines indicated that 90 % of the fruits travelled 20 m or less from the stem of the mother tree and a summary of other dipterocarp studies compiled by the same authors shows that most fruits landed within 30 m or, at most, 40 m. This compares with a dispersal distance within 2-3 m of the crown perimeter for a heavy, wingless seed such as Quercus crispula in Japan and a distance of over 60-90 m for 5 % of the light, winged seeds of Betula ermannii downwind from a belt of mother trees left in a logged over area.

Fleshy edible fruits and arillate seeds, on the other hand, encourage dispersal by birds or mammals. When such fruits or seeds are eaten by animals, the seeds, protected by the hard seedcoat or endocarp, often pass unharmed through the digestive tract and are deposited in the faeces at a considerable distance from the place where they were consumed. In many cases the digestive juices actually assist subsequent germination through softening of the hard seedcoat. In Africa the hornbill is a highly efficient dispersal agent for seeds of Maesopsis eminii. Sometimes the process is so effective as to be an embarrassment.

Free-ranging goats eat the pods of Prosopis in some countries and spread the seeds indiscriminately over large areas; excellent germination and the aggressive pioneering qualities of the young plants may then render this genus a dangerous weed-tree. Coralling of the goats and collection of the seeds for use under strictly controlled management can overcome the problem. In other cases the fruit is eaten while the stones or seeds are rejected, but the animal may carry the fruit some distance from the parent before dropping the seeds. Rodents remove and store nuts or seeds; many are subsequently eaten but some may escape to germinate in the new situation.

Wind and animals are the most important agents of dispersal, but dispersal by water is common in some riverine species and large and heavy fruits are distributed to some extent by gravity on steep slopes.

Gymnosperm Seed Development

Gymnosperm ovules have certain characteristics in common with angiosperm ovules, but there are a number of differences. There is normally a single protective integument which in a typical female cone is partially fused to the ovuliferous scale carrying the paired ovules. Within the integument is the nucellus which at fertilisation, as in angiosperms, is clearly separated from the integument only in the region of the micropyle. Meiosis within the nucellus, followed by mitotic cell divisions, leads to the formation of a multicellular haploid tissue—the female gametophyte. By the time of fertilisation it has developed much further than the 8-nucleate embryo sac in the angiosperms and has largely displaced the nucellus. At its micropylar end it is differentiated into one to many archegonia, each of which contains a large egg cell.

At fertilisation the pollen tube releases two male nuclei into an archegonium, one of which unites with the egg nucleus. The resulting zygote later develops into the new diploid embryo. The second male nucleus aborts in Pinus but may fertilise a second archegonium in other genera e.g. Cupressus. It never unites with female polar nuclei to form a triploid tissue analogous to the endosperm of angiosperms; this type of tissue is unknown in gymnosperm seeds. For detailed descriptions of gymnosperm embryogeny, readers are referred to the specialised literature.

The mature seed consists of some or all of the following: (1) The seedcoat or testa developed from the integument, diploid from the female parent. (2) The diploid perisperm, developed from the nucellus. In most species this is absorbed by the female gametophyte and has disappeared by the time the seed is ripe, but it is still recognisable as a distinct tissue in e.g. Pinus pinea. (3) The haploid female gametophytic tissue which serves as a food storage organ to nourish the embryo. Its function is the same as the endosperm in angiosperms and it is frequently called by that name, though this usage has been deprecated. (4) The embryo, with the same parts of radicle, cotyledons, plumule and hypocotyl as in angiosperms.

The number of cotyledons varies between and within genera, being up to 18 in Pinus, compared with the constant two in the dicotyledons which comprise the great majority of angiosperm trees. The essential constituents of embryo, protective covering and food storage tissue are present in all gymnosperm, as in all angiosperm, seeds. More than one archegonium may be fertilised within a single ovule, but in the great majority of cases only one embryo per seed develops to maturity. Polyembryony does occur but is uncommon in most genera.

Gymnosperm Fruit Development

After fertilisation the female cone which is typical of several important gymnosperm genera e.g. Pinus, Picea, Pseudotsuga, Araucaria increases in size and weight, in moisture content and accumulated food reserves. As the cones approach maturity, the moisture content decreases again, accumulated food reserves move from cone to seed and the cone becomes more or less woody.

In Pinus a thin membranous flake becomes detached from the ovuliferous scale and adheres to the ripe seed, forming a wing. In Juniperus the cone scales grow together to form a fleshy berry-like fruit, while in Podocarpus and Taxus each singly borne seed becomes partly enclosed in a brightly coloured cup, the aril. The woody cone is, however, the most characteristic type of fruit in gymnosperms.

As in angiosperms, there is wide variation in the interval between flowering and seed maturity and dispersal. The total period between pollination and cone maturity is usually about two years in this genus; among tropical pines, average periods are 23 months in Pinus kesiya, 18-21 months in P. oocarpa. In Agathis robusta it takes 16 months from pollination to cone maturity, in Araucaria cunninghamii up to 24 months, in Araucaria hunsteinii 21-24 months. In several temperate genera development is completed within a single season e.g. in 5 months in Pseudotsuga menziesii. Development of unpollinated female cones with fully formed but usually empty seeds occurs in a number of gymnosperm genera. Parthenocarpy is common in Abies, Juniperus, Larix, Picea, Taxus and Thuja. It is rare in pines.

Seed Dispersal in Gymnosperms

In the typical gymnosperm cone, ripening and drying of cone and seed causes the cone scales to open and release the seeds. Dispersal is by wind, assisted by the presence of seed wings in some genera e.g. Pinus. In some species of pine, the "closed-cone pines" e.g. P. radiata, there is usually an interval of months or years between ripening of cone and seed and the opening of the cone to release the seeds. In a few cases, such as the interior provenances of Pinus contorta, the cones open only when subjected to the heat of occasional fierce forest fires. On the other hand the cones of Abies and Araucaria disintegrate readily on the tree within a few weeks of ripening.

Seed dispersal by animals is less common, but the "berries" of Juniperus and the fleshy fruits of Podocarpus are examples. In addition seeds of temperate conifers are collected and stored by rodents and some may germinate before being eaten.

Seed Germination

At one extreme, certain species of mangrove are viviparous, the seeds germinating before they separate from the parent. At the other extreme, seed of some species may remain dormant but alive for many years, capable of germinating if an event occurs to break the dormancy state. Between the viviparous and the deeply dormant seed types occur many types of seed which are capable of germination soon after seed shed provided that environmental conditions are suitable.

Just as fertilisation initiates the transformation of the ovule into the ripe seed, so does germination transform the embryo within the seed into the independent seedling. For the purpose of laboratory testing, germination is defined as the emergence and development from the seed embryo of those essential structures which are indicative of the seed's capacity to produce a normal plant under favourable conditions.

At maturity and seed shed many seeds have lost the greater part of the moisture which they contained in earlier stages. For example the embryo and female gametophyte of Pinus lambertiana contain as much as 50 % moisture content (fresh weight basis) shortly after fertilisation, but by the time of natural seed dispersal the moisture content of the embryo is reduced to 23 % and of the female gametophyte to 38 %. Reduced metabolic activity is associated with the drying of the seed, so that the embryo is in a temporarily resting or quiescent state, which in non-dormant seeds can be easily reactivated by suitable conditions. These conditions are (1) Adequate moisture (2) Favourable temperatures (3) Adequate gas exchange and, for some species, (4) Light. There is considerable variation between species in the optimum levels of the different factors and there is frequently an interaction between them.

Germination consists of three overlapping processes: (1) Absorption of water mainly by imbibition, causing a swelling of the seed and eventual splitting of the seedcoat, (2) Enzymatic activity and increased respiration and assimilation rates which signal the use of stored food and translocation to growing regions, (3) Cell enlargement and divisions resulting in emergence of radicle and plumule.

In most seeds the radicle of the embryo is close to the micropyle, where absorption of water is easier and quicker than through the seedcoat. As the radicle swells, it exerts pressure on the seedcoat which commonly splits first at this point to free the radicle. This gives rise to the primary root which grows down into the soil and soon produces lateral roots. Subsequent stages depend on whether the species exhibits epigeal germination e.g. Pinus—the hypocotyl elongates and the cotyledons are lifted above ground—or hypogeal germination e.g. Quercus—the hypocotyl is undeveloped and the cotyledons remain on or in the ground.

In hypogeal germination the cotyledons can have only a food storage function, or a haustorial function (in species in which food is stored in the endosperm e.g. palms, Scorodocarpus), while in epigeal germination they may also perform a valuable photosynthetic function during early growth of the seedling.

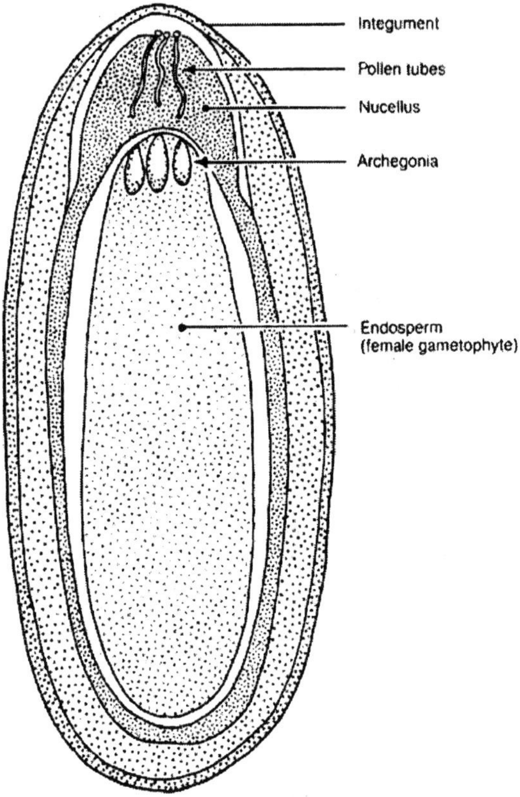

Figure 3. Longitudinal section through an ovule of Pinus during the period of pollen tube development preceding fertilisation.

In epigeal germination anchoring of the young plant by the radicle is followed by rapid elongation of the hypocotyl which arches upwards above the soil surface and then straightens; simultaneously the cotyledons and plumule are exposed, to which the seedcoat may or may not still be attached. The plumule then develops into the primary shoot and photosynthetic leaves. In the subtype "durian germination" the hypocotyl elongates but the cotyledons are shed while still enclosed within the seedcoat (e.g. Durio zibethinus, Strombosia javanica).

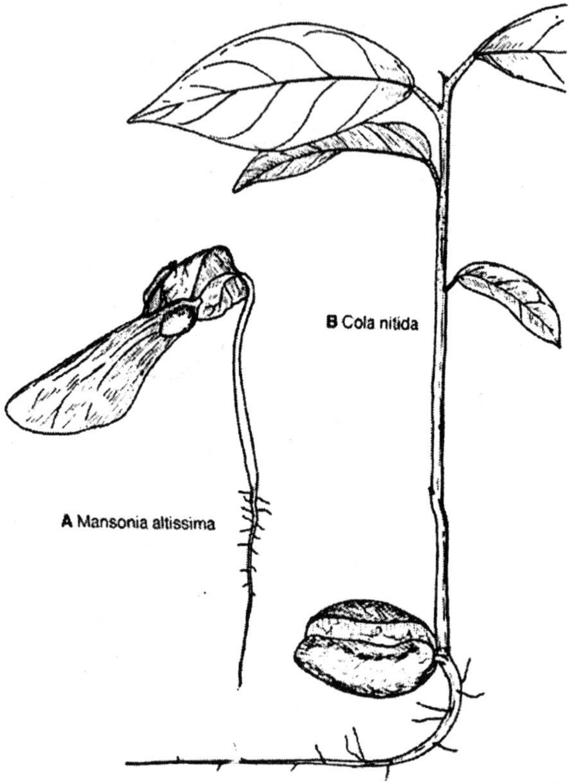

Figure 4. Examples of germination in two West African Sterculiaceae. (A) Epigeal in Mansonia altissima. (B) Hypogeal in Cola nitida.

In hypogeal germination, the cotyledons remain in situ underground or on the ground while elongation takes place in the plumule. In the subtype "semihypogeal germination" the cotyledons are exposed but remain on the ground. The two main types, epigeal and hypogeal and the two subtypes, durian and semihypogeal are the result of the four possible combinations of two independent variables: hypocotyl elongated or not and cotyledons exposed or not. All four combinations occur in the humid tropics.

Even in non-dormant seeds there is considerable variation, between species and individuals, in the speed of germination, from a few days to several weeks; much of this is due to different rates of imbibition in the first stage. Many tropical rain forest species, with high moisture content and permeable seedcoats at seed fall, must germinate within a few weeks. If they fail to find suitable conditions soon, they lose viability and die.

Dormancy

The term "dormancy" refers to a condition in a viable seed which prevents it from germinating when supplied with the factors normally considered adequate for germination—suitable temperature, moisture and gaseous environment. A viable seed is defined as one which can germinate under favourable conditions, providing any dormancy that may be present is removed.

Dormancy in nature serves to protect the seed from conditions which are temporarily suitable for germination but which quickly revert to conditions too harsh for survival of the tender young seedling. Thus a seedcoat relatively impermeable to moisture prevents germination during isolated showers in the middle of a long dry season, while permitting it during a sustained rainy season. In the cool temperate zone the type of embryo dormancy which can be removed only through exposure to low temperatures facilitates subsequent germination in spring, while preventing it in autumn, when the resulting seedling would be unlikely to survive the winter.

The strength of dormancy has been observed to vary according to latitude and provenance, and from year to year even in seed from the same parent. There is also differential dormancy within the same species and seedlot, so that germination is staggered over a more or less extended period of time. In the Malaysian woody flora, about 50 % of species complete germination within six weeks or less, so that the differential is not very great, but in some species such as the hard-seeded leguminous Parkia javanica the germination period may extend from 1 week after sowing for the first seed to two years for the last. Differential dormancy and staggered germination insures against the entire seed crop being destroyed by a single climatic catastrophe or pest.

In nature a number of external factors may work, more or less slowly, to end seedcoat dormancy. These include alternate heating and cooling, alternate wetting and drying, fire and the activities of animals, soil organisms, fungi, termites and other insects. Dormancy due to embryo immaturity will come to an end if the embryo is given time and suitable conditions in which to mature after seed shed.

The exact mechanisms of physiological dormancy of the embryo, and of the processes which can terminate it, have been widely investigated but underlying causes are still little understood. There is good evidence that growth-promoting hormones, of which gibberellin is a well known example, and growth-inhibiting hormones interact in the maintenance or breaking of dormancy.

In temperate climates the balance between inhibitors and growth-promoters is altered by a combination of low temperature and high moisture maintained over a period of time which varies from species to species. This combination is provided naturally during winter, the season least suitable for growth. It can trigger biochemical changes in the embryo which lead to the breaking of dormancy, the initiation of embryo metabolism and growth, and the subsequent germination of the seedling.

Research on the physiology of tropical tree species has, unfortunately, been only a fraction of that done on temperate species. There is no reason to suppose that the combination of low temperature and high moisture—the technique of "stratification" when applied artificially—would have any effect on tropical seeds possessing embryo dormancy (if they exist). In the dry tropics, on the contrary, a combination of conditions typical of the season least favourable for growth—high temperature and low moisture—would seem more likely to trigger the breaking of embryo dormancy and lead to germination in the following rainy season. In fact, seedcoat dormancy by itself appears to provide adequate protection for species of the dry tropics.

From the forester's point of view dormancy has some disadvantages. Delayed and irregular germination in the nursery is a serious constraint on efficient nursery management. Much research has therefore gone into devising effective artificial treatments to remove dormancy, in order to ensure that the seeds germinate quickly and evenly in the nursery beds.

On the other hand, dormancy confers certain advantages. Not only does it improve the chances of survival in nature, as mentioned previously, but it preserves the seed against temporarily unsuitable conditions such as may occur during the period between seed collection and storage. High quality orthodox but non-dormant seeds, dried to the appropriate moisture content and stored at the correct temperature, should have as long a life in storage as dormant seeds, but dormancy provides an insurance against the loss of viability during transport and processing which can easily occur in non-dormant seeds in less than ideal conditions.

Hazards of Seed Production

Both the quantity and the quality of seed crops may be greatly affected by external factors. Climatic factors can affect the abundance of flowering and thus indirectly of seed production. There is some evidence that above-average temperatures and a modest degree of moisture stress in spring and early summer can induce abundant formation of flower buds in temperate regions. In Nigeria good seed years occur in Triplochiton scleroxylon following a particularly dry (30 % or less of average rainfall) August, the month when there is a diminution of rainfall between the heavy early and heavy late rains.

More extreme cases of unseasonal climate usually reduce the crop of flowers or fruits. Late spring frosts in temperate regions kill flowers or young fruits and abnormally high temperatures or drought may have a similar effect. Even if death and premature shed of whole fruits does not occur, a proportion of the seeds may abort later. Mechanical destruction of flowers or fruits may be caused by exceptionally high winds or hailstorms. Continuous rain at the time of pollen dispersal has a particularly adverse effect cn the amount of seed set, whether pollination is by wind or insects.

Tectona flowers during the rainy season and this may account for the low average rate of fertilisation of 1 - 3 % reported from Thailand cver the period 1967 - 72. Rain discourages flight of the pollinating insects as well as washing off pollen grains from the stigma before they germinate. Continuous wet weather during the pollen dispersal season is considered the main factor responsible for the characteristically poor seed crops of Pinus merkusii in Indonesia and Malaysia. Tamari reported that over 90 % of Dipterocarp flowers in Malaysia failed to develop into fruits.

Birds, mammals, insects, fungi and bacteria all do damage in flowering as well as fruiting stages. Insects are probably responsible for the most serious losses in the greatest number of species. For example, the weevil Apion ghanaense destroys a large part of the flowers and seeds of Triplochiton each year. The larvae of Pagyda salvaris may destroy as much as 90 % of the flower buds of Tectona in some years. Two species of the Bruchid genus Amblycerus can destroy many seeds of Cordia alliodora, but damage can be reduced by collecting the seeds three weeks before natural seedfall. The weevil Nanophyes sp. may attack up to 60 % of the seeds of Terminalia ivorensis.

Cone worms of the genus Dioryctria have been known to damage 60 % of maturing cones and seeds of Pinus elliottii and P. palustris in the southern USA and the same genus can also cause severe damage on Pinus merkusii seeds in the Philippines. Larvae of Agathiphaga, a genus of moths, may destroy over 50 % of the seeds in cones of several species of Agathis in Queensland and the western Pacific islands. Seeds of many dry area species of Acacia and Prosopis suffer serious damage from Bruchid larvae. Birds and mammals, especially squirrels, may consume considerable numbers of seeds in some years, although they also perform a useful service in seed dispersal. Losses from pests and diseases do not normally have a serious effect in years of abundant seed production, but in years when flowering is poor for climatic reasons they can convert a light crop into complete failure.

Bibliography

FAO, "Developing seed security strategies and programmes for food security in developing countries", *Internal document of the Seed and Plant Genetic Resources Service of FAO*, Rome, Italy, 1997.

Hong, T.D., Linington, S. and Ellis, R.H., "Compendium of Information on Seed Storage Behaviour", Volume I A–H. Royal Botanic Gardens, Kew, UK, 1998.

ISTA, *Seed Science and Technology— Rules*. Vol. 24. ISTA, Zurich, 1996.

Jacobs, M.R., *Eucalypts for Planting, FAO,* Forestry Series No. 11. FAO, Rome, 1981.

Justice, O.L. and Bass, L.N., "Principles and practices of seed storage", *Agric. Handbook,* 506.Washington DC: U.S. Department of Agriculture, 1979.

Khumalo, D., "Minimizing the risk associated with on-farm seed development in: Proceedings of the Regional Workshop on Improved On-Farm Seed production for SADC countries", *FAO*, Italy, 1993.

Larinde, M., "Improving seed supply to small islands countries", Lessors from the CARICOM countries, *Paper presented at the Workshop on Improved Seed Supply Strategies for Pacific Islands Countries held in Nadi Fiyi,* Poggenpoel, P.V. 1978.

Ndjeunga, J., "Local village seed systems and pearl millet seed quality in Niger.", *Experimental Agriculture,* 2002.

Poulsen, K and Stubsgaard, F., "Three methods for mechanical scarification of hardcoated seed", Technical Note No. 27. Danida Forest Seed Centre, Humlebaek, Denmark, 1995.

Roberts, E.H., "Predicting the storage life of seed", *Seed Science and Technology,* 1973.

Stubsgaard, F. and Poulsen, K.M., "Seed moisture and drying principles", *Lecture Note* No. C-5. Danida Forest Seed Centre, Humlebaek, Denmark, 1995.

Teramura A.H., J.H. Sullivan, J. Lydon, "Effects of UV-B radiation on soybean yield and seed quality: a six-year field study", Physiologia Planatarum, 1990.

Thomson, L.A.J., "Collecting woody perennials", In: Guarino, L., Ramanatha Rao, V. and Reid, R. eds, *Collecting Plant Genetic Diversity,* Technical Guidelines. IPGRI, Rome, 1995.

Turnbull, J.W., "Assessment of seed crops and the timing of seed collections", *In: Report on FAO/DANIDA Training Course on Forest Seed Collection and Handling,* Vol. 2, FAO, Rome, 1975.

Willan, R.L. (compiler), "A guide to forest seed handling", *FAO Forestry Paper* 20/2. FAO, Rome, 1985.

Index